THE
INTELLIGEN
OF THE
COSMOS

D0205030

"From the time when the conscious universe was a preposterous notion to today, when it's a cutting-edge idea full of promise for your future, Ervin Laszlo has been its staunch champion. Now he and his colleagues press the argument with even more urgency and depth. I concur that consciousness must be regarded as the ground state of existence. Only when we begin to see ourselves embedded in a reality where mind, body, ecology, planet Earth, and the cosmos are all expressions of the same consciousness will we truly know who we are."

DEEPAK CHOPRA, COAUTHOR OF *YOU ARE THE UNIVERSE*

"Ervin Laszlo is without any doubt one of the greatest scientists and philosophers of our time. His seminal concepts of the Akashic holofield and the connectivity hypothesis represent an extraordinary intellectual achievement that will be seen in the future as a critical landmark in the history of science. Laszlo succeeded in something previously deemed impossible—creating a 'map of everything' that dissolves the boundaries between natural sciences, psychology, philosophy, and spirituality. In his latest book, *The Intelligence of the Cosmos,* Laszlo summarizes in a clear and captivating style his profound understanding of the nature of reality and human nature that has emerged from six decades of his unrelenting search for truth. His work might become the cornerstone of the new comprehensive scientific worldview of the twenty-first century."

STANISLAV GROF, M.D., AUTHOR OF *BEYOND THE BRAIN,*
PSYCHOLOGY OF THE FUTURE, AND *THE COSMIC GAME*

"Professor Ervin Laszlo, author of the 1987 world-changing classic *Evolution: The Grand Synthesis,* again gives us a great work for the future, *The Intelligence of the Cosmos: Why Are We Here? New Answers from the Frontiers of Science.* For the entire past century science has inched steadily toward a synthesis of physics, cosmology, and biology. Increasingly cosmology looks more and more like evolutionary biology, and intelligence is discovered in the tiniest nooks and crannies of the universe. Now comes the bold step in which Laszlo points out the obvious, showing us that the universe *is* a living intelligent being. This book is science at its most engaging and absolute best!"

ALLAN COMBS, PH.D., PROFESSOR AT THE CALIFORNIA INSTITUTE
FOR INTEGRAL STUDIES, AUTHOR OF *THE RADIANCE OF BEING*
AND *CONSCIOUSNESS EXPLAINED BETTER,*
AND COEDITOR OF *THOMAS BERRY: DREAMER OF THE EARTH*

"In 1897 the French painter Paul Gauguin titled one of his masterpieces *Where Do We Come From? What Are We? Where Are We Going?* The painting ended up in the Museum of Fine Arts in Boston. But those questions, and many others brilliantly embraced in this remarkable work, *The Intelligence of the Cosmos,* continue to haunt every aspect of our lives. Ervin Laszlo and colleagues have gently and demonstrably entered a domain of that luminous haunting, that "supercoherence" which they reveal to be rooted to the quintessential mysteries of all life forms: life's origins, destiny, and seemingly incomprehensible, flickering instant of Being. It is a book like no other and offers a remarkable and hopeful context for better navigating these colossally challenging times—for biodiversity, for humanity, and for something indescribable shaping our entire universe and ourselves. Indeed, this may be the most important book ever written on the subject of consciousness, against the backdrop of a starry night and the alluring earth beneath our feet."

MICHAEL CHARLES TOBIAS, PH.D.,
PRESIDENT OF DANCING STAR FOUNDATION,
A GLOBAL ECOLOGIST, ANTHROPOLOGIST,
HISTORIAN, EXPLORER, AUTHOR, AND FILMMAKER

"I highly recommend this timely cutting-edge material from Ervin Laszlo, as it brings together science-based evidence in a way to understand the transformational revolution and new insights into consciousness and the evolution of humankind taking place now!"

KEVIN MOORE, HOST OF *THE MOORE SHOW*

THE
INTELLIGENCE
OF THE
COSMOS

WHY ARE WE HERE?

NEW ANSWERS FROM THE
FRONTIERS OF SCIENCE

ERVIN LASZLO

Inner Traditions
Rochester, Vermont • Toronto, Canada

Inner Traditions
One Park Street
Rochester, Vermont 05767
www.InnerTraditions.com

Library of Congress Cataloging-in-Publication Data

Names: Laszlo, Ervin, 1932– editor.
Title: The intelligence of the cosmos : Why are we here? : New answers from the frontiers of science / [edited by] Ervin Laszlo.
Description: Rochester, Vermont : Inner Traditions, [2017] | Includes bibliographical references and index.
Identifiers: LCCN 2017033609 (print) | LCCN 2017015674 (e-book) | ISBN 9781620557310 (pbk.) |ISBN 9781620557327 (e-book)
Subjects: LCSH: Cosmology. | Life. | Meaning (Philosophy) | Consciousness. | Physics.
Classification: LCC BD523 (print) | LCC BD523 .I58 2017 (e-book) | DDC 113—dc23
LC record available at https://lccn.loc.gov/2017015674

Printed and bound in the United States by Versa Press, Inc.

10 9 8 7 6 5 4 3 2

Text design by Priscilla Baker and layout by Debbie Glogover
This book was typeset in Garamond Premier Pro with Gill Sans MT Pro, Helvetica Neue LT Std, ITC Legacy Sans Std for display fonts

Cover image courtesy of iStock.

To send correspondence to the author of this book, mail a first-class letter to the author c/o Inner Traditions • Bear & Company, One Park Street, Rochester, VT 05767, and we will forward the communication, or contact the author directly at **ervin@ervinlaszlo.it**.

MY CREDO

Eight Cardinal Propositions

1. *The cosmos* is an infinite and eternal intelligence.

2. The infinite and eternal intelligence brought into being a finite domain of space and time: *the universe*.

3. The things we observe, or infer from observation, are *clusters of vibration in the universe*, in-formed by the intelligence of the cosmos.

4. At different frequencies and wavelengths, clusters of in-formed vibration are perceived as *structures of matter*, as *individual consciousness*, and as *transcendental intuition*.

5. Clusters of vibration perceived as structures of matter and as individual consciousness evolve in space and time. Structures of matter (matter-like clusters of vibration) evolve intermittently: they periodically de-cohere and reconfigure. Individual consciousnesses (mind-like clusters of vibration) evolve continuously, through incarnate phases in association with structures of matter and discarnate phases beyond matter and beyond space and time.

6. Structures of matter evolve toward supercoherence, and individual consciousnesses evolve toward oneness with and love for all things in space and time.

7. The purpose of the evolution of clusters of vibration in the universe is the reception and transmission of the intelligence of the cosmos into the universe.

8. The ultimate purpose of *human* existence is to consciously foster and further the transmission of the unifying, embracing, and all-encompassing intelligence of the cosmos into the universe.

Contents

⊕

PART ONE

Perennial Questions and the New Answers

PART TWO

The Meaning of the New Answers for Our Life and Times

Acknowledgments

It is the pleasure and privilege of this writer to thank his insightful and wise friends and colleagues for their contribution to this book. They have brought their personal experience, insight, and accumulated wisdom to bear on the questions posed here. They joined the Laszlo Institute of New Paradigm Research in the quest to spell out new paradigm answers to the fundamental questions of our life and existence—including who we are and why we are here. I wish to express my heartfelt thanks and deep appreciation to Emanuel Kuntzelman, president of Greenheart International; John Audette, CEO of the Eternea Organization; Maria Sagi, science director of the Club of Budapest; Garry Jacobs, CEO of the World Academy of Art and Science; as well as Kingsley Dennis and Shamik Desai of the Laszlo Institute of New Paradigm Research.

My sons, Christopher and Alexander, and my wife, Carita, have been, as always, steadfast and loving supporters in bringing this new project to fruition, and so has my agent and good friend, William Gladstone.

It has been a pleasure to work again with my publisher, Ehud Sperling, and his outstanding team at Inner Traditions, in particular Jon Graham, John Hays, Jeanie Levitan, Eliza Burns, and

Annie Downey. Their care, expertise, and attention to every detail have been a source of constant satisfaction to me, and assurance that my fondly nurtured "baby" is launched into the world in the best possible shape.

Montescudaio, Italy

Summer 2017

Foreword

Jane Goodall, Ph.D., DBE

Who are we? and *Why are we here?* are questions that have preoccupied humans for thousands of years. Is our universe, including our own little planet, the result of mindless chance? Are we humans simply one more entity that has gradually emerged throughout the long course of evolution on planet Earth? Just one more creature of flesh and blood, albeit one with an unusually highly developed brain? Or is there more to us than that—a spiritual dimension, a consciousness, that exists independently of our physical body, independently of our brain, though all is intricately connected? Are we the only creature on Earth to question who we are and why we are here, what is the purpose of our lives?

I have spent many years in the forest studying the behavior of our closest living relatives, the chimpanzees. Biologically we differ from them in the composition of our DNA by only just over 1 percent (with the main difference, I am told, being in the expression of the genes). And there are remarkable similarities too in the composition of blood and immune system and anatomy of the brain. There is now proof that chimpanzees are far more intelligent than mainstream science used to believe; they can solve complex problems, learn four hundred or more signs of American Sign Language (ASL), and use these to engage in meaningful exchanges with humans and even other apes. And many other mammalian species are far more intelligent than most scientists, until very recently, were prepared to admit so that, finally, there is much interest in investigating the intelligent behavior of birds, octopuses, bees—and even communication between trees.

Chimpanzees and many other animals, we now conclude, are sentient beings, capable of feeling joy and sadness, anger, grief, depression, and other emotions. Certainly they feel pain. Like us, chimpanzees have a dark side and are capable of violence and even a kind of primitive warfare. Like us, they also show compassion and empathy. They have a definite sense of self: they can recognize themselves in mirrors. They can understand the wants and needs of others. They have a theory of mind. And it is possible that they have what I can only describe as a sense of awe at the wonder of nature. For example, they perform impressive and rhythmic displays at the base of magnificent waterfalls deep in the forest and then sit, watching as the water falls down and down and then flows past them and away. What is it that is always coming, always going, always here? If they could but discuss these feelings might this not result in the emergence of an animistic religion: worship of the powerful forces of nature, of water, sun, moon . . . ?

It is the explosive development of our intellect, possibly triggered in part by our ability to communicate, using words (spoken or written) about things that are not present, to make plans for the distant future, above all to discuss ideas and problems, that sets us apart from other animals. Some chimpanzees and other animals love to paint—that is they make marks on paper with different colors and even distinctive patterns—but even the most talented of them could not produce a Rembrandt or a van Gogh, or even the smiling or angry stick figures drawn by small children in kindergarten. Nor could they formulate a theory of relativity or unravel the human genome. And it is inconceivable they could puzzle over the kind of questions asked by Laszlo in this book.

Of course, along with increasing intelligence and a deeper understanding of consciousness, we have also developed a heightened sense of morality. A sense of good and bad behavior. My understanding of the natural world has taught me that there is much that, by our moral standards, is cruel—"nature red in tooth and claw," as Tennyson famously wrote. But we cannot equate a carnivore killing and eating its prey, even eating the unfortunate victim while it is still alive, with a human being who deliberately inflicts torture, physical or mental, on another sentient being. The animal is merely obeying its nature. Our heightened moral values and superior intellect put us in a different position. We can cause harm to another with full understanding that our actions will cause pain, fear, distress. Only this, I believe, can be described as evil. And this we must fight, in ourselves and others, as we move through life.

In fact, our purpose is to do good, to honor the Golden Rule by doing unto others as we would have them do unto us, an admonition that is present, in a similar form, in just about all the great religions. And it is easy to understand why this should be so. But we are also admonished to "love our neighbors as ourselves," and for a long time I queried this: how could I love myself, with all my imperfections, my sometimes selfish or unkind behavior? Until finally I realized that the "self" I was to love was the flame of pure spirit that is in each of us, linked to the One Consciousness of the universe. And that which is loved can grow.

It is very clear that during the billions of years of the evolution of life on planet Earth there has been a gradual trend to create ever more complex and coherent forms of life. And thanks to the work of brilliant scientific minds it can now be shown that this could not have come about by chance. Mathematical physicist Fred Hoyle said that the probability that new species would emerge through a chance mutation of their genes is comparable to the probability that a hurricane blowing through a scrapyard would assemble a working airplane. And so we must accept that there is an Intelligence driving the process, that the universe and life on Earth are inspired and in-formed by an unknown and unknowable Creator, a Supreme Being, a Great Spiritual Power— or the Intelligence, as this book shows, that is intrinsic to the cosmos.

We cannot but be awed when we think of the succession of brilliant minds that have, over the ages, come to the conclusion that there is, indeed, a Creator, One who caused the Big Bang that created the universe. It is a fact that there are rules that govern the universe and that these are mathematical rules that have led many to believe in a Supreme Intelligence. Richard Feynman, a Nobel Prize winner for quantum electrodynamics, said, "Why nature is mathematical is a mystery. . . . The fact that there are rules at all is a kind of miracle."

Francis Collins was director of the Human Genome Project (and is currently director of the National Institutes of Health). He became increasingly amazed by the complexity of our DNA—which in essence, he explains, is a three-billion-lettered program that informs each cell to act in a certain way, a full instruction manual comprised of chemicals that instruct how our bodies are to develop. Surely, he wrote in *The Language of God,* only an awesome Intelligence could have caused this complex manual to wind up in each human cell . . .

Until relatively recently philosophers and scientists were not able to prove that what they believed was, indeed, true. Mainstream science still debunks the idea of a supreme spiritual force in the universe, but Laszlo points out that mainstream science has not grasped the real significance of the complexity of the laws and the phenomena of nature. He has drawn on the most up-to-date findings in the fields of cosmology, physics, and the life sciences to support the idea that there is Purpose underlying the creation of the universe, that consciousness has evolved through scientifically demonstrable processes, and that mind and spirit can be shown to exist beyond and independently of the physical body. He discusses near-death experiences and has assembled impressive examples of out-of-body and other transcendental experiences. (I had my own transcendental experience in contacting the mind of one who was physically dead.) And he discusses Eastern religions that believe in the reincarnation of gradually evolving consciousness into successive physical flesh-and-blood bodies.

I was brought up in a Christian family and took it for granted that God existed. This belief was strengthened during my years in the forest as I learned about the interconnectedness of life-forms, the amazing mystery of the natural world. And I became a part of it. I learned, first-hand, that I and nature were one consciousness. In fact, in some strange way, I think I always understood. I was still at school when I wrote the following:

The Old Wisdom

When the night wind makes the pine trees creak
And the pale clouds glide across the dark sky,
Go out, my child, go out and seek
Your soul: the Eternal I.

For all the grasses rustling at your feet
And every flaming star that glitters high
above you, close up and meet
in you: the Eternal I.

Yes, my child, go out into the world; walk slow
And silent, comprehending all, and by and by
Your soul, the Universe, will know
Itself: the Eternal I.

And then, a few years later after I had a wonderful moment with a duck, this became the last verse:

The lovely dunes; the setting sun;
The duck—and I;
One Spirit moving timelessly
Beneath the sky.

After I left the forest and began traveling around the world talking about the need to protect the environment, wildlife, and biodiversity, I met more and more people, especially high school and university students, who seemed to have lost hope. They were mostly just apathetic, but some were depressed and some were angry, even violent. They thought about what we were doing to the planet—the deforestation, pollution, loss of biodiversity, shrinking freshwater supplies, the effect of chemicals used in agribusiness and industry, our damaging effects producing climate change. There is the refugee crisis. And everywhere we find racial discrimination, hatred, exploitation, greed, and corruption. There is ethnic violence, terrorism, and war. The world is a mess.

No wonder the young people told me that we had compromised their future—we have. But when they said there was nothing they could do about it, I could not agree. I believe there is a window of time during which, if enough of us work to create a new mind-set, we can heal some of the wounds we have inflicted, and at least slow down environmental deterioration and climate change.

Laszlo tells us that we need to be concerned with the evolution of the consciousness of a critical mass of people in every part of the world. Most people's consciousness, he told me, is evolved enough to know that they need change, but not evolved enough to know what is the change they need. What is needed is a critical mass of people who become more coherent with each other and with the world around them. And that is exactly why, in 1991, I created my Roots & Shoots program. It was to create a critical mass of young people who understand that there has been a loss of wisdom, a disconnect between the clever brain and the loving heart. To bring together young people who become empowered by working on projects (that *they* choose) to make this world a better place for people, animals, and the environment. Who learn that every individual matters and has a role to play. Who understand that if each one of us thinks about the consequences of the little choices we make each day—what we buy, eat, wear—the collective result of millions of ethical choices will move us toward a better world. And, perhaps most importantly, who, by communicating with each other, sharing ideas and results, are breaking down the barriers that we build between people of different nations, cultures, religions . . . between us and the natural world. Who learn that no matter the color of our skin, if we are wounded our blood is the same, and we all have feelings of joy and sadness, despair and pain. This movement is now in one hundred countries with some 1,500 active groups of all ages. These groups communicate with each other, share ideas—about respect, love, and compassion. Many who have been part of the movement, who have been empowered, are already out there, as parents, teachers, lawyers, business people, politicians, and others.

Of course, it is perfectly possible to go through life helping others, showing love and compassion, without a belief in God—but how much more rewarding to know that you are fulfilling the purpose for which you came into the world! And that is why I am so grateful to you, Ervin, for writing this book, *The Intelligence of the Cosmos*. It is so important, coming at this critical time. Young people are searching for meaning in life. They desperately need to find hope for the future. And this book, based on cutting-edge science that *proves* the existence of a Master Mind behind the creation of the universe and provides meaning for our lives, is just exactly what the young people of today need.

PART ONE

Perennial Questions and the New Answers

1

Who Are We?

Who are we? And why are we here?

Despite their apparent simplicity, these are perhaps the most important and challenging questions ever asked. "Know thyself," said the oracle at Delphi, and thinking people have been trying to do so for millennia. Today, at the cutting edge where modern science meets timeless spirituality, we can give answers that are more trustworthy than any of the answers oracles and humans could give before.

We need to start with the basics. Who are we? Are we part of the world in which we live, or are we beings apart from and perhaps above the world? We may possess unique qualities and features, but science assumes that we, as all living things, are part of the world in which we came to be. But what is the true nature of that world? If we are to know who we are, we need to know what the world is. Remarkably, the answer to this query is becoming well-founded and reliable—even though it is very different from what most people think it is.

A New Concept of the World

The world, according to cutting-edge science, is not an ensemble of bits and pieces of matter. As astrophysicist James Jeans said over a hundred years ago, the world is more like a big thought than like a big rock.

The concept of a thoughtlike world is familiar from the history of speculation on fundamental questions. Today, more and more philosophers, scientists, and intuitive people in all walks of life question that the world would be just what is presented to our senses. They realize that the classical answer—that the universe would be the arena for the motion of insensible matter in passive space and indifferently flowing time—is not the last word. The old concept, based on Newton's classical mechanics, has run its course. There is a new concept of what the world truly is. It is not an ensemble of separate bits of matter moving in accordance with mechanistic laws, but an intrinsically whole system where all things are connected in ways that transcend the previously known bounds of space and time.

In the new concept, the things that furnish the world are not pieces of matter. Surprisingly (or perhaps not so surprisingly, because this has been an age-old intuition) they are basically *vibration*. The vibrations that make up the world we observe are not random but highly ordered: they are *coherent*. Their order and coherence tell us that they are not the result of mere chance. To use a term suggested by physicist David Bohm, the vibrations are "in-formed." (Bohm used this term to explain how the deep dimension of the cosmos beyond space and time—he called it "the implicate order"—affects the perceived dimension in space and in time: the "explicate order.") Everything in the world is spontaneously and effectively in-formed by a deep dimension, ourselves and our brain and mind included. In the last count, we are in-formed clusters of vibration in space and time, interacting and coevolving with other clusters both locally, here and now, and nonlocally, throughout the universe.

In one form or another, the concept of the world as in-formed vibration has been part of the wisdom of the great spiritual traditions. It was present in the Sanskrit concept of Akasha, where it meant the fifth, and deepest, dimension or element of the world, beyond yet encompassing the four elemental dimensions of earth, air, fire, and water. It was sometimes used in the sense of "sky" or "atmosphere." Later it came to be seen as an ethereal field underlying the observed world.

The intuition of the Akasha was present in the Vedic texts of India as early as 5000 BCE. In the Vedas its function was identified with *shabda,* the first vibration, the first ripple that makes up our universe, and also with *spanda,* described as "vibration/movement of consciousness." The contemporary Indian scholar I. K. Taimni wrote,

> There is . . . a mysterious integrated state of vibration from which all possible kinds of vibrations can be derived by a process of differentiation. That is called *N.da* in Sanskrit. It is a vibration in a medium . . . which may be translated as "space" in English. But . . . it is not mere empty space but space which, though apparently empty, contains within itself an infinite amount of potential energy.

This formerly esoteric notion is now sustained and substantiated in science. In quantum physics, observations and calculations reveal that at the ultrasmall dimension, space is not empty and smooth. It is "grainy," filled with waves and vibrations. When physicists descend to the ultrasmall dimension, they do not find anything that could be called matter. What they find are waves and clusters of standing or propagating vibrations.

Previously, scientists assumed that it is matter that vibrates. There is a ground substance that vibrates, and that substance consists of matter particles and assemblies of matter particles. The world is material, and vibration is the way matter behaves. But the contrary turned out to be the case. There is no ground substance. The world is a set of variously integrated clusters of vibration, and matter is just the way the vibrations appear to observers.

The great physicist Max Planck said it clearly. In one of his last lectures in Florence he noted, "As a man who has devoted his whole life to the most clear-headed science, to the study of matter, I can tell you as a result of my research about atoms this much: There is no matter as such. All matter originates and exists only by virtue of a force which brings the particle of an atom to vibration and holds this most minute solar system of the atom together." And, Planck added, behind the force that holds the vibrations of the atomic nucleus together "we must assume the existence of a conscious and intelligent mind."* That intelligence, he said, is the matrix of all matter.

Planck was not alone in reviving the concept of the world as wave and vibration. Two years prior to Planck's pronouncement, the maverick genius Nicola Tesla advised that if you want to know the secrets of the universe, you should think in terms of energy, frequency, and vibration.

Today we can say clearly and with assurance that the materialist concept of the world is obsolete. The new sciences tell us that it is not from bits of matter but from clusters of highly ordered *in-formed* vibration that the things we find in space and time are built.

The affirmation "the world is vibration" begs a further question. If the world is vibration, what is the world the vibration of? *What is it that vibrates?* It could not be the ground substance of the universe, for we have no independent evidence for the existence of such a substance. It could also not be matter, because we have no independent evidence for the existence of matter either. *It could be space.* As the classical wisdom schools have maintained, there is more to space than a location free of matter. Space is not empty and passive; it is filled and dynamic. Dynamically filled space could vibrate.

Today we know that what physicists familiarly (but misleadingly) call the quantum vacuum is not a vacuum at all; it is not empty space. It is a *plenum,* a space filled with vibrations and forces of various kinds, some known, such as the electromagnetic, gravitational, and nuclear fields, and others yet to be defined, including fields and forces associated

---

Das Wesen der Materie [The Nature of Matter], speech in Florence, Italy, 1944. Archiv zur Geschichte der Max Planck Gesellschaft, Abt. Va, Rep. 11 Planck, Nr. 1797.

with dark energy and dark matter. It is a universal subquantum field, the domain this writer called the Akashic field.*

The Akashic subquantum field is researched in science under various names. It is the grand unified field of particle physics, the zero point field of electrodynamics, the universal quantum field of quantum mechanics, and the implicate order of David Bohm. Everything that emerges and persists in the world is a cluster of vibration in that field. The human body and bodies of all shapes and sizes, from atoms to galaxies, are clusters of vibration in the Akashic field of the universe.

A New Concept of Consciousness

What about mind? If the world is vibration, is also mind and consciousness a form of vibration? Or on the contrary, are all vibrations, the observed world, a manifestation of mind?

Although it is true that when all is said and done all we know is our consciousness, it is also true that we do not know our own consciousness, not to mention the consciousness of anyone else. We do not know what consciousness really is or how it is related to the brain. *Since our consciousness is the basis of our identity, we do not know who we really are.* Are we a body that generates the stream of sensations we call consciousness, or are we a consciousness associated with a body that displays it? Do we *have* consciousness, or *are* we consciousness? Consciousness could be a kind of illusion, a set of sensations produced by the workings of our brain. But it could also be that our body is a vehicle, a transmitter of a consciousness that is the basic reality of the world. The world could be material, and mind could be an illusion. Or the world could be consciousness, and the materiality of the world could be the illusion.

Both of these possibilities have been explored in the history of philosophy, and today we are a step closer than before to understanding

---

*See the author's *Science and the Akashic Field* (Rochester, Vt.: Inner Traditions, 2004); 2nd ed. (2007).

which of them is true. There are important insights emerging at the expanding frontiers where physical science joins consciousness research.

On the basis of a growing series of observations and experiments, a new consensus is emerging. It is that "my" consciousness is not just *my* consciousness, meaning the consciousness produced by my brain, any more than a program transmitted over the air would be a program produced by my TV set. Just like a program broadcast over the air continues to exist when my TV set is turned off, my consciousness continues to exist when my brain is turned off.

Consciousness is a real element in the real world. The brain and body do not produce it; they display it. And it does not cease when life in the body does. Consciousness is a reflection, a projection, a manifestation of the intelligence that "in-forms" the world.

Mystics and shamans have known that this is true for millennia, and artists and spiritual people know it to this day. Its rediscovery at the leading edge of science augurs a profound shift in our view of the world. It overcomes the answer the now outdated materialist science gives to the question regarding the nature of mind: the answer according to which consciousness is an epiphenomenon, a product or by-product of the workings of the brain. In that case, the brain would be like an electricity-generating turbine. The turbine is material, while the current it generates is not (or not strictly) material. In the same way, the brain could be material, even if the consciousness it generates proves to be something that is not quite material.

On first sight, this makes good sense. On a second look, however, the materialist concept encounters major problems. First, a conceptual problem. How could a material brain give rise to a truly immaterial stream of sensations? How could anything that is material produce anything immaterial? In modern consciousness research this is known as the "hard problem." It has no reasonable answer. As researchers point out, we do not have the slightest idea how "matter" could produce "mind." One is a measurable entity with properties such as hardness, extension, force, and the like, and the other is an ineffable series of sensations with no definite location in space and an ephemeral presence in time.

Fortunately, the hard problem does not need to be solved: it is not a real problem. There is another possibility: mind is a real element in the real world and is not *produced* by the brain; it is *manifested* and *displayed* by the brain.

Mind beyond Brain: Evidence for a New Concept of Consciousness

If mind is a real element in the real world only manifested rather than produced by the brain, it can also exist without the brain. There is evidence that mind does exist on occasion beyond the brain: surprisingly, conscious experience seems possible in the absence of a functioning brain. There are cases—the near-death experience (NDE) is the paradigm case—where consciousness persists when brain function is impaired, or even halted.

Thousands of observations and experiments show that people whose brain stopped working but then regained normal functioning can experience consciousness during the time they are without a functioning brain. This cannot be accounted for on the premises of the production theory: if there is no working brain, there cannot be consciousness. Yet there are cases of consciousness appearing beyond the living and working brain, and some of these cases are not easy to dismiss as mere imagination.

A striking NDE was recounted by a young woman named Pamela. Hers has been just one among scores of NDEs;* it is cited here to illustrate that such experiences exist, and can be documented.

Pamela died on May 29, 2010, at the age of fifty-three. But for hours she was effectively dead on the operating table nineteen years ear-

---

*For a more extensive sampling see Ervin Laszlo with Anthony Peake from *The Immortal Mind* (Rochester, Vt.: Inner Traditions, 2014).

lier. Her near-demise was induced by a surgical team attempting to remove an aneurism in her brain stem.

After the operation, when her brain and body returned to normal functioning, Pamela described in detail what had taken place in the operating theater. She recalled among other things the music that was playing ("Hotel California" by the Eagles). She described a whole series of conversations among the medical team. She reported having watched the opening of her skull by the surgeon from a position above him and described in detail the "Midas Rex" bone-cutting device and the distinct sound it made.

About ninety minutes into the operation, she saw her body from the outside and felt herself being pulled out of it and into a tunnel of light. And she heard the bone saw activate, even though there were specially designed speakers in each of her ears that shut out all external sounds. The speakers themselves were broadcasting audible clicks in order to confirm that there was no activity in her brain stem. Moreover, she had been given a general anesthetic that should have assured that she was fully unconscious. Pamela should not have been able either to see or to hear anything.

It appears that consciousness is not, or not entirely, tied to the living brain. In addition to NDEs, there are cases in which consciousness is detached from the brain in regard to its location. In these cases consciousness originates above the eyes and the head, or near the ceiling, or above the roof. These are the out-of-body experiences: OBEs.

There are OBEs where congenitally blind people have visual awareness. They describe their surroundings in considerable detail and with remarkable accuracy. What the blind experience is not restored eyesight, because they are aware of things that are shielded from their eyes or are beyond the range of normal eyesight. Consciousness researcher Kenneth Ring called these experiences "transcendental awareness."

Visual awareness in the blind joins a growing repertory of experiences collected and researched by Stanislav Grof: "transcendental experiences."

As Grof found, these beyond-the-brain and beyond-here-and-now experiences are widespread—more widespread than anyone would have suspected even a few years ago.

There are also reports of ADEs, after-death experiences. Thousands of psychic mediums claim to have channeled the conscious experience of deceased people, and some of these reports are not easy to dismiss as mere imagination. One of the most robust of these reports has come from Bertrand Russell, the renowned English philosopher. Lord Russell was a skeptic, an outspoken debunker of esoteric phenomena, including the survival of the mind or soul beyond the body. He once wrote, "I believe that when I die I shall rot, and nothing of my ego will survive." Yet after he died he conveyed the following message to the medium Rosemary Brown.

> You may not believe that it is I, Bertrand Arthur William Russell, who am saying these things, and perhaps there is no conclusive proof that I can offer through this somewhat restrictive medium. Those with an ear to hear may catch the echo of my voice in my phrases, the tenor of my tongue in my tautology; those who do not wish to hear will no doubt conjure up a whole table of tricks to disprove my retrospective rhetoric.
>
> . . . After breathing my last breath in my mortal body, I found myself in some sort of extension of existence that held no parallel as far as I could estimate, in the material dimension I had recently experienced. I observed that I was occupying a body predominantly bearing similarities to the physical one I had vacated forever; but this new body in which I now resided seemed virtually weightless and very volatile, and able to move in any direction with the minimum of effort. I began to think I was dreaming and would awaken all too soon in that old world, of which I had become somewhat weary to find myself imprisoned once more in that ageing form which encased a brain that had waxed weary also and did not always want to think when *I* wanted to think. . . .

Several times in my life [Lord Russell continued] I had thought I was about to die; several times I had resigned myself with the best will that I could muster to ceasing to be. The idea of B.R. no longer inhabiting the world did not trouble me unduly. Befitting, I thought, to give the chap (myself) a decent burial and let him be. Now here I was, still the same I, with the capacities to think and observe sharpened to an incredible degree. I felt earth-life suddenly seemed very unreal almost as it had never happened. It took me quite a long while to understand that feeling until I realized at last that matter is certainly illusory although it does exist in actuality; the material world seemed now nothing more than a seething, changing, restless sea of indeterminable density and volume.

This report "from beyond" appears hardly credible, were it not that it is supported by other ADEs. One of the most striking and difficult to dismiss of these ADEs is the case of a deceased chess grand master who played a game with a living grand master.*

Wolfgang Eisenbeiss, an amateur chess player, engaged the medium Robert Rollans to transmit the moves of a game to be played with Viktor Korchnoi, the world's third-ranking grand master. His opponent was to be a player whom Rollans was to find in his trance state. Eisenbeiss gave Rollans a list of deceased grand masters and asked him to contact them and ask who would be willing to play. Rollans entered his state of trance and did so. On June 15, 1985, the former grand master Geza Maroczy responded and said that he was available. Maroczy was the third-ranking grand master in the year 1900. He was born in 1870 and died in 1951 at the age of eighty-one. Rollans reported that Maroczy responded to his invitation as follows.

I will be at your disposal in this peculiar game of chess for two reasons. First, because I also want to do something to aid mankind

---

*For details see Laszlo with Peake, *The Immortal Mind*.

living on Earth to become convinced that death does not end everything, but instead the mind is separated from the physical body and comes up to us in a new world, where individual life continues to manifest itself in a new unknown dimension. Second, being a Hungarian patriot, I want to guide the eyes of the world into the direction of my beloved Hungary.

Korchnoi and Maroczy began a game that was frequently interrupted due to Korchnoi's poor health and numerous travels. It lasted seven years and eight months. Speaking through Robert Rollans, Maroczy gave his moves in the standard form: for example, "5. A3 – Bxc3+"; Korchnoi gave his own moves to Rollans in the same form, but by ordinary communication. Every move was analyzed and recorded. It turned out that the game was played at the grand-master level and that it exhibited the style for which Maroczy was famous. It ended on February 11, 1993, when at move forty-eight Maroczy resigned. Subsequent analysis showed that it was a wise decision: five moves later Korchnoi would have achieved checkmate.

In this case the medium Rollans channeled information he did not possess in his ordinary state of consciousness. And this information was so expert and precise that it is extremely unlikely that any person Rollans could have contacted would have possessed it.

There are also firsthand testimonies of consciousness without a functioning brain. The well-known Harvard neurosurgeon Eben Alexander, who was just as insistently skeptical about consciousness beyond the brain as Lord Russell had been, gave a detailed account of his conscious experience during the seven days he spent in deep coma. In the condition in which he found himself, conscious experience, he previously said, is completely excluded. Yet his experience—which he described in detail in several articles and three bestselling books—was so clear and convincing that it has changed his mind. Consciousness, he is now claiming, can exist beyond the brain.

The above-cited cases illustrate that there is remarkable, and on occasion remarkably robust, evidence that consciousness is not confined to the living brain. Although this evidence is widespread, it is not widely known. There are still people, including scientists, who refuse to take cognizance of it. This is not surprising, given that the evidence is anomalous for the dominant world concept. Those who strongly disbelieve that such phenomena exist, not only refuse to *consider* evidence to the contrary, they often fail to *perceive* evidence to the contrary.

Nonetheless, the view that consciousness is a fundamental element in the world is gaining recognition. The Manifesto of the Summit on Post-Materialist Science, Spirituality and Society (Tucson, Arizona, 2015) declared: "Mind represents an aspect of reality as primordial as the physical world. Mind is fundamental in the universe, i.e., it cannot be derived from matter and reduced to anything more basic."

An In-Formed World

It appears that consciousness is not limited to the individual brain and body; it is a fundamental element in the universe. The universe, as we now know, is not a domain of matter moving in passive space and indifferently flowing time; it is a sea of coherent vibrations. These vibrations give us the phenomena of physical realities such as quanta, atoms, solar systems, and galaxies, and they also give us the phenomena of nonphysical realities: mind and consciousness.

The affirmation that physical vibrations give rise to nonphysical mind phenomena is not just another version of the "hard question" of consciousness research—the problem of how something as material as the brain can give rise to something as immaterial as consciousness. This however is a pseudo-problem, since clusters of vibration do not *produce* the phenomana of consciousness; they *manifest* and *display* them. The cosmos that gave birth to the universe is fundamentally an intelligence, and that intelligence is manifest in—"in-forms"—all phenomena.

The hard question evaporates if we realize that the physical world is a domain, a segment, and hence a manifestation, of the intelligence of the cosmos. The vibrations that produce the phenomena of physical and nonphysical phenomena are part of the reality of the world, a world that is in-formed by, and manifests, the intelligence that is not only "of" the cosmos, but *is* the cosmos.

The vibrations that manifest the cosmic intelligence are not physical entities in the classical sense of the term. They have a physical as well as a nonphysical aspect. Viewed from the outside, every cluster of vibration is a physical phenomenon, a pattern of vibration in space and time. But viewed from the inside—from the perspective of the given cluster—it is a perception, an awareness, a *"feeling"* of the world in and by that cluster. This internal, seemingly subjective but objectively real aspect is a fundamental feature of the universe. It is the consciousness-aspect, one that did not emerge in the course of time but was present when this universe was born in the wider reality of the cosmos.

The bottom line is that the phenomena that appear as consciousness were not created or produced by the clusters of vibration in which they appear: they are *manifested* by them. Consciousness is, and has always been, present in all clusters, from quanta to galaxies. It is not limited to those with a complex brain and nervous system, even though the level of its manifestation corresponds to the level of evolution of the clusters that manifest it—chimpanzees manifest a more advanced consciousness than mice.

The clusters that manifest consciousness in the universe were not, and cannot have been, the product of chance—chance alone does not explain the presence of any cluster of vibration more complex than a hydrogen atom. It does not explain the presence of the simplest of biological organism. Calculations show that a random mixing of the proteins that constitute that DNA of the common fruit fly would have taken more time than was available since the Big Bang. There is something in the universe—a mind, a principle, or an intelligence—that orders and "in-forms" the phenomena that structure and hold together the phenomenawe observe in the world. The medium of this "in-formation" is the ensemble of the laws that govern events in the universe. Einstein wrote: "Everyone who is seriously involved in the pursuit of science becomes convinced that a spirit is manifest in the laws of the universe—a spirit vastly superior to that of man, and one in the face of which we with our modern powers must feel humble."* Planck came to the same conclusion. He said that we must assume the presence of an intelligence even behind the vibrations that constitute the nuclei of atoms. This must be a universal intelligence. It is the intelligence that not only holds together the proton and the neutron in the nuclei of atoms but holds together atoms in molecules—and molecules in the multi-molecular structures that are the physical objects of the observed universe.

We have reason to maintain that a cosmic intelligence is manifested in our consciousness as well. Evidently, not everything that appears in our everyday awareness is a mark of a cosmic intelligence; our everyday consciousness is mainly furnished by sights and sounds, textures, odors, and tastes conveyed by our bodily senses. They result from our brain's decoding of vibrations in our environment. Our eyes pick up a narrow band of vibrations in the electromagnetic field and transform them into shapes and colors of determinate brightness; our ears pick up a likewise narrow if somewhat wider band of vibrations in the air and transform them into sounds of specific pitch and intensity. But sensory experience, while it produces the principal content of our experience of the world, is not the totality of our conscious awareness. Beyond the data of the senses there are images and intuitions, experiences and feeling tones that are not decoding of ambient vibrations by our eyes and ears: they

---

*There are various versions of this sentence, which comes from a letter Einstein wrote in 1936 in his native German. It is worthy of note that the German word *Geist,* translated here as "spirit," is much more basic and embracing than the esoteric notion of spirit: *Geist* also means "mind" and "intelligence." In the context of its implications for science, Einstein's sentence tells us that a mind or intelligence must be assumed to be present behind the laws of nature.

are trans-sensory, "transcendental" elements of consciousness. They emerge into prominence when the everyday operations of the brain are slowed, or are shut down.

Transcendental experiences are a standard feature of NDEs, but they surface also in sleep, and in the hypnogogic states between sleep and wakefulness. They appear in traumatic, uplifting, or otherwise life-transforming experiences. And, if the reports channeled by psychic mediums are true, they emerge following the physical death of the body. That they do stands to reason. Following the demise of the brain, the sensory elements of consciousness are withdrawn, and the transcendental elements alone dominate.

Consciousness beyond the brain was previously an esoteric assertion, but it now has a scientific foundation. The clusters of vibration that convey our awareness of the world include clusters that convey transcendental experiences. These clusters are not fundamentally different from those that convey everyday phenomena; they merely occupy a different frequency domain. Clusters at the frequency corresponding to the square of the velocity of light produce the phenomenon of pure energy, per Einstein's mass-energy equation $E = mc^2$. Clusters in the frequency domain of the speed of light produce the sensation of light, and those at the frequency domain of the living cell—about 150 megahertz—give us the perception of the everyday world.

There are vibrations of an extreme low frequency as well. These low-frequency, long-wavelength vibrations are likely to convey elements of the intelligence that in-forms our body and our brain—and in-forms the universe.

Indirect support for this hypothesis comes from the measurements of the EEG frequencies generated by the brain. In everyday states of awareness, pronounced excitations of the brain trigger EEG waves in the gamma domain, a frequency range of up to 100 hertz. The everyday world appears in the beta range, which is between 12 and 30 hertz. Deeply relaxed or meditative states occur in the low-frequency band known as alpha: 7.5–12 hertz. Transcendental experiences occur mainly in the still lower band of theta, between 4 and 7 hertz.

Below theta, we have the super-low range of delta, between near-zero and 4 hertz. Delta waves are normally produced by the brain only in deep sleep. But there are exceptions. The brain of some psychic healers has been known to descend into this region while they are engaged in healing. And spiritual leaders in deep meditation also function in this super-low region, although they are not in the unconscious state of deep sleep.

We need to revise the widespread assumption that nonordinary "spiritual" experiences occur at a high-frequency domain. They do not occur *above* the frequency of everyday experiences, but *below* it.

Ordinary, bodily sense-based experience is not the totality of human experience, only a small part of it. We are a cluster of vibration that manifests phenomena of a far wider range than the narrow segment that gives us the sensory data the old paradigm told us is the totality of our experience of the world.

2
Why Are We Here?

Why are we here? is perhaps the most challenging question ever asked. It needs to be posed today in light of the evidence coming to light at the frontiers of science.

Hamlet said that to be or not to be is the question, but he did not say what the meaning of our existence is if we choose "to be." Is there a purpose for our being in the world? Purpose is a controversial concept; conservative scientists question that it is meaningful even to ask about it. The things that furnish the universe, they say, are the product of random processes that occur without rhyme or reason, and not for any purpose. We are the products of such processes, and there is no higher or deeper purpose for our being in the world.

Spiritual and religious people do not agree; they believe that there is a purpose for our existence in the world: it is to obey the will of God—or the principles and decrees of Yahweh, Mohammed, Tao, Brahman, or the Great Spirit.

In the context of the paradigm emerging in science, it is meaningful to ask about the purpose of being in the world. The world, we have seen, is an interconnected, intrinsically whole system in-formed by something we can recognize as an intelligence—the intelligence that, in the final count, is the cosmos itself. We are not just flesh-and-bone chunks of matter in an indifferent universe, but beings with mind and consciousness in a vibrating quantum domain. There could be purpose for the existence of such beings; it is unlikely that they would be here merely by chance.

The conservative mainstream of the scientific community is reluctant to accept that there would be an intelligence underlying things and processes in the universe, and hence that there would be anything like purpose for the existence of things, including human beings. Cosmologist Steven Weinberg said, "I believe there is no point that can be discovered by the methods of science. I believe what we have found so far—an impersonal universe which is not particularly directed towards human beings—is what we are going to find. And that when we find the ultimate laws of nature they will have a chilling, cold, impersonal quality about them."*

Mainstream scientists hold on to the belief that the universe is the outcome of a long chain of basically random interactions. The things that result from them have not been purposively created; they just happen to come about. Physicist Nassim Haramein explained,

> The fundamental axioms and basic assumptions at the root of physical theories . . . presume that evolutionary systems emerge from random interactions initiated by a single "miraculous" event providing all of the appropriate conditions to produce our current observable universe, and our state of existence in it. This event, typically described as a "Big Bang," astonishingly is thought to have produced all of the forces and constants of physical law and eventually biological interactions under random functions.[†]

Thomas Kuhn, the originator of the concept of scientific paradigms, recognized that there is a consistently unfolding order in nature,

---

*Interview cited in Laszlo, *Science and the Akashic Field* (Rochester, Vt.: Inner Traditions, 2007).

†Nassim Haramein, in Ervin Laszlo's *What Is Reality? The New Map of Cosmos and Consciousness* (New York: SelectBooks, 2016).

but did not allow that this would be evolution toward a given goal and thus that it would have a deeper purpose. He wrote,

> The developmental process has been an evolution from primitive beginnings—a process whose successive stages are characterized by an increasingly detailed and refined understanding of nature. But nothing makes it a process of evolution toward anything. . . . The entire process may have occurred, as we now suppose biological evolution did, without benefit of a set goal, a permanent fixed scientific truth, of which each stage in the development of scientific knowledge is an improved exemplar."*

Evolution, according to Kuhn, is an intelligible but not a purposive process, it is not evolution *toward* anything. However, an unbiased assessment of the way evolution unfolds leads to a different conclusion. The long-term processes of evolution are not random and they are not reversible: they manifest a distinct directionality. They drive or tend toward particular states and conditions. These states and conditions are discoverable and, in general terms at least, they can be described. Whatever their exact nature and destination may be, they suggest the presence of purpose in nature.

Indications of Purpose
in the Direction of Evolution

There can be purpose for a nonrandom process unfolding in a particular direction. We can discover the nature of that purpose by noting the direction in which the process unfolds. This means discovering the direction exhibited by the evolution of the things that exist in the universe. This is what we undertake here. We first examine the direction

---

*Thomas, Kuhn, *The Structure of Scientific Revolutions* (University of Chicago Press, 1962).

in which natural systems (as opposed to artificial, man-made systems) evolve in the universe, and then examine the direction underlying the evolution of the nonmaterial yet fundamental phenomena we recognize as mind and consciousness.

Direction in the Evolution of Natural Systems

As we have seen, the entities that populate space and time are clusters of vibration of various dimensions of size and complexity. They are integrated clusters: "systems" coming about in the course of evolutionary processes. Particles and atoms, molecules and cells, as well as planets and solar systems are integrated clusters: complex coherent systems, the products of natural processes structuring the vibrations that emerged in the wake of the Big Bang. Why and how they have come about has puzzled philosophers as well as scientists.

Abrahamic, Hermetic, Vedic, and Daoist thinkers ascribed the emergence of the things that furnish the universe to supernatural agency, while scientists searched for natural causes. Philosopher Henri Bergson postulated an *elan vital* that counters the trend toward the degradation of energy in living systems, and biologist Hans Driesch called for a counterentropic drive he termed *entelechy.* Teilhard de Chardin and Erich Jantsch invoked the notion of *syntony,* and other investigators spoke of *syntropy* as the force behind the evolution of complex and coherent systems. Whatever name we give to the factor that structures the vibrations that emerged in the universe, it is clear that it is an orienting factor: a cosmic goal, or *Telos,* in the basic Greek sense of the word.

The Findings

Not only do complex and coherent systems evolve in the universe, the universe itself is a complex and coherent system. Already in the middle of the twentieth century, Sir Arthur Eddington and Paul Dirac noted curious "coincidences" concerning the parameters of the universe.

The ratio of the electric force to the gravitational force is known: it is approximately 10^{40}. This is an enormous number, but it is nearly the same as that which defines the ratio between the size of the whole universe and the size of the minute quantum particles that appear in it. This is surprising, since the ratio of the electric force to the gravitational force should be unchanging (given that these forces are constant), whereas the ratio of the size of the universe to the size of elementary particles should be changing (since the universe is expanding). In his "large number hypothesis," Dirac speculated that the agreement between these ratios, one variable the other not, is more than random coincidence. But if so, either the universe is not expanding or the force of gravitation varies proportionately to its expansion.

Later cosmologist Menas Kafatos showed that many of the ratios among the parameters of the universe can be interpreted on the one hand in terms of the relationship between the masses of elementary particles and the total number of nucleons, and on the other in reference to the relationship between the gravitational constant, the charge of the electron, Planck's constant, and the speed of light.

Astonishingly, the mass of elementary particles, the number of particles, and the forces between them all display harmonic ratios. Even the microwave background radiation—the remnant of the Big Bang—is unexpectedly coherent: it is dominated by a large peak followed by smaller harmonic peaks. The series of peaks ends at the longest wavelength, which physicist Lee Smolin termed R. When R is divided by the speed of light (R/c) we get the length of time independent estimates have shown to be the age of the universe. When we divide in turn the speed of light by the constant R, we get a frequency that equates to one cycle over the age of the universe. And when R is squared and divided by the speed of light (R^2/c) we get the measure of the acceleration in the expansion of the galaxies. These are astounding "coincidences." They tell us that the universe is a coherent system as a whole. Its parameters are finely tuned to one another, and together they are coherent with its overall dimensions.

The coherence of the parameters of the universe is extremely precise: variations of the order of one-billionth of the value of some constants (such as the mass of elementary particles, the speed of light, the rate of expansion of galaxies, and some two dozen others) would not have produced stable atoms and stable interaction among the atoms. Already a minute variation of some of the physical constants would have precluded the evolution of the coherent systems we call living. The fact is that living systems are astonishingly coherent. The human body is made up of 10^{14} cells, and each cell produces ten thousand bioelectrochemical reactions every second. Every twenty-four hours 10^{12} cells die and are replaced. In our body, molecules, cells, and organ systems resonate at the same or at compatible frequencies and interact at various speeds, ranging from the slow (among hormones and peripheral nerve fibers) to the very fast (along the Ranvier rings of myelin-shielded nerves). The interactions are precisely correlated, involving quantum-type "entanglements" in addition to classical physical-biological interactions.

The universe is highly coherent in itself, and it has brought forth highly coherent systems. Many systems possess a remarkable measure of intrinsic as well as extrinsic coherence. "Intrinsic coherence" means that the parts that make up the systems are finely tuned together, so that every element is responsive to every other element. "Extrinsic coherence" in turn means that the systems are coherently connected to other systems around them. Evolution in the universe exhibits a drive or tendency toward creating intrinsically as well as extrinsically coherent systems.

The evolution of complex and coherent systems calls for an explanation. Chance, even if involving a large number of systems over large time scales, cannot account for the facts: the search-space of the possible configuration of the elements that make up the systems is so vast that random trial and error would have greatly exceeded the available time frames.

There were 13.8 billion years available for the evolution of natural systems in the universe and more than four billion years for the evolution of living organisms on this planet. This span of time, although enormous, is insufficient to explain the presence of the highly coherent systems we now encounter. The probability that even the simplest biological organisms that populate the biosphere would have come about through a random shuffling of their elements is negligible. The DNA-mRNA-tRNA-rRNA transcription and translation system, basic to living systems, is so complex and precise that it is astronomically improbable that living organisms could have evolved through a chance assembly of their genes. As Jane Goodall noted, the probability that new species would emerge through a chance mutation of their genes is comparable to the probability that a hurricane blowing through a scrapyard assembles an airplane.

In the mainstream of science, a series of fortunate coincidences is cited as the explanation of the evolution of life on this planet. Earth is in a fortunate location in the galaxy, neither too far nor too close to its sun, a main-sequence G2 dwarf star. It has the right atmosphere and the right amount of water for producing and sustaining life, it has the right mass, and it occupies a nearly circular orbit. It has an oxygen-nitrogen-rich atmosphere and a moderate rate of rotation. There is liquid water on its surface, and a correct ratio between water and landmass. Its surface temperature fluctuates within the narrow range required for life. It is also at the right distance from the center of the galaxy and is protected from asteroids by giant gas planets. In this position the sun's heliosphere protects Earth's surface from cosmic rays and pressures lethal for biological systems, and the planet's own magnetosphere protects it from dangerously high energies emanating from the heliosphere.

Earth is indeed in a fortunate location: in a so-called "Goldilocks zone."* Such a location is not pure serendipity. Some two thousand

---

*This refers to the fairy tale of Goldilocks and the three bears. Goldilocks found the house of the three bears and in their absence tried the porridge and the beds and discovered that some are too hot or too big, and others too small and too cold. Finally she found those that are just right—like Earth did in the Milky Way galaxy.

"exoplanets" have been identified, planets orbiting stars other than our sun, and scientists working with the Kepler space telescope detected several thousand other stars that may have exoplanets. On the average, each star in the Milky Way galaxy has at least one planet, and one in five "sunlike" stars is likely to have an Earth-size planet in the Goldilocks zone. With two hundred billion stars in our Milky Way galaxy, there could be eleven billion Earth-size planets in the Goldilocks zone in this galaxy alone—and there are 10^{22} to 10^{24} galaxies in the universe—this is highly probable because the evolution of coherent systems is not due solely to fortunate conditions.

Organic molecules, the basic elements of life, appear to have been synthesized under a surprisingly wide range of conditions. A team of astrophysicists headed by Sun Kwok and Yong Zhang at the University of Hong Kong found 130 macromolecules present even in the vicinity of active stars. They include glycine, an amino acid, and ethylene glycol, the compound associated with the formation of the sugar molecules necessary for life. Their presence suggests that they were ejected in the course of the stars' thermal and chemical evolution.

Organic molecules were discovered in interstellar clouds as well. The incidence of the most complex of these molecules, isopropyl cyanide, has been reported in 2014 by a team of researchers headed by Arnaud Belloche at the Max Planck Institute for Radio Astronomy. Its branching carbon structure is similar to that of the amino acids that form the basis of proteins on this planet.

Further evidence that the evolution of coherent systems is coded into the basic processes of the universe came with the discovery that primordial DNA self-creates spontaneously. The spontaneous self-assembly of DNA fragments a few nanometers in length into liquid crystals drives the formation of chemical bonds and creates chains of DNA. The self-organizing properties of DNA-like molecular fragments over billions of years may have produced the first DNA-like molecular chains on Earth, the same as on Goldilocks-zone planets elsewhere in the universe.

The Buildup of Coherent Complexity

The evolution of coherent systems, cosmologists believe, began in the wake of the singularity known as the Big Bang. But we have reason to believe that the Big Bang was not the beginning of physical reality in the world; it was the beginning only of this particular part of physical reality: our universe. Other universes may have existed prior to the Big Bang. There are findings that suggest that there was a universe, and possibly many universes, prior to the Big Bang. For example, we see galaxies near the edge of the observable universe as they were 13 billion years ago (it took as long for light from that distance to reach us), but many of these galaxies appear to be fully evolved, with stars ranging in age from one to ten billion years. That, however, places their origins up to 23 billion years before our time. Moreover, there are supermassive black holes [SMHs] at the center of this and other galaxies, and some of them have an estimated solar mass of 10^9, yet the known rates at which stars are captured in black holes would have taken far longer than fourteen billion years to produce structures of this dimension. It appears that the Big Bang marked the birth of our universe in the context of a "multiverse" that included other universes. If all universes were in-formed by the same cosmic intelligence as ours, they are likely to follow a basically similar evolutionary trajectory as our own universe.

Our universe, and all universes that exist in the multiverse, were born as the energies released by their "local" Big Bang polarized the segment of the cosmic Akashic field that became their point of origin. This brought this hitherto quiescent segment of the field (the so-called Minkowski vacuum) into vibration, creating Planck-scale "ripples": leptons (electrons, muons, tau particles, neutrinos), mesons (pions), as well as hadrons (baryons, including protons and neutrons). In the course of time they acquired structure. Some of these "ripples" clustered into more embracing clusters: atoms, molecules, and multi-molecular structures. On the astronomical scale, stars and stellar systems, galaxies, and galactic clusters came into being.

The laws of nature are instructions, precise algorithms, for the evolution of coherent systems in the spacetime domain of our universe. On the physical level the crucial instruction is the Pauli exclusion principle. The principle states that no two electrons orbiting the nucleus of an atom can be at the same quantum state at the same time. Electrons entering the gravitational zone of the nucleus are excluded from orbits that are already occupied and are distributed into other orbits, filling up the energy shells that surround the nucleus. Due to their exclusion, the particles captured by the gravitational field of the nucleus assemble into coherent and complex structures. As a result, the physical world is not a heap or conglomeration of particles, but a domain of coherent systems. It is built of the atoms that populate the periodic table of the elements and of the structures that form as the atoms bond with other atoms in molecules and crystals. Molecules form multimolecular structures, and these are templates for still more complex chemical and biological systems.

Other laws of nature join the exclusion principle in making the universe into a domain of coherent systems. Coherent systems are inevitably complex. A higher form of organization in a complex system does not just repeat the structure on the lower levels, but adds novelty, while repeating key patterns that remain invariant. Multilevel systems in nature are fractal (self-similar) ensembles of cooperative parts, where all parts cooperate in maintaining the system in the physically improbable state far from thermal and chemical equilibrium. Such a system embodies long-range interactions that optimize connection among its elements, safeguarding and enhancing the coherence of the system as a whole. It is multilevel and thus complex.

Similarly to Einstein, Deepak Chopra and Menas Kafatos attributed the laws and regularities that create coherent systems in nature to a cosmic intelligence.* They defined how such a cosmic consciousness "behaves"—how it creates coherent systems. They group the structure- and coherence-creating laws of nature under a few basic headings:

> ***Complementarity.*** Positive and negative, yin and yang, as all opposites balance each other without abolishing or diminishing one another.

> ***Creative interactivity.*** As diverse elements interact, not just more of the same results; new forms and functions come into existence.

> ***Evolution.*** The old is the basis for creating the new, and when the new is created, it integrates the old without destroying or neutralizing it.

> ***Veiled nonlocality.*** Locally separate things and events are nonlocally joined together at a deeper and less evident level.

> ***Cosmic censorship.*** Everything is connected with everything else, yet local perspectives remain valid: the interconnections of the whole are kept from view.

> ***Recursion.*** All parts and elements of the whole share patterns and forms that mirror and repeat each other at successively deeper levels.

Due to the action of these and related laws, complexity emerges and grows in the universe. Evolution builds physical entities into complex and coherent physicochemical structures, and these into still more complex and coherent biological systems.

---

*Deepak Chopra and Menas Kafatos, *You Are the Universe* (New York: Harmony Books, 2017).

Coherence in the systems cannot grow indefinitely. Beyond a critical level of size and diversity, the systems become unstable, de-cohering into their individually stable components. When the limits of stability have been reached, the systems either de-cohere into stable components or join and create systems at a higher level of complexity. As a result, there is a progression from level to level of structure and complexity in nature: from the atomic to the molecular, from the molecular to the multimolecular, from the multimolecular to the cellular and multicellular, and from there to the ecological and biospherical. The overall direction in the evolution of the natural systems is toward higher levels of coherence and complexity.

How complexity builds in biological systems has been described by biologist Bruce Lipton. He noted,

> First, a new organism evolves (e.g. a bacterium, a protist, a multicellular animal or plant). Evolution then leads to the creation of the smartest individual version of that new organism.
>
> Second, when physical limitations on increasing awareness are reached, the individual organisms collectively assemble into structure communities to further maximize their intelligence. This process is complete when the cellular community evolves into a "new" single organism, which starts the process over again.*

This view is a radical departure from the classical theory of biological evolution. According to the Darwinian concept, the evolution of species is toward fitness, where fitness means the optimum—that is, optimally stable and enduring—adaptation of species to the environment. But if fitness were the goal of evolution, the biosphere would be populated mainly by blue-green algae, amoebae, and other unicellular, colonial, and simple multicellular organisms. Many of these species achieved a

---

*Bruce Lipton, *A New Model of Evolution: Realizing Our Positive Future.* Paper contributed to the Club of Budapest's Twentieth Anniversary Conference. See www.theclubof budapest.com.

nearly perfect adaptation to their environment, and nothing short of volcanic eruptions, sudden climatic change, and natural catastrophe could lead to their extinction.

Yet the biosphere is not primarily populated by superfit and superstable simple organisms. Species evolve not just toward, but also beyond, the range of optimum fitness. They explore the environment for all niches that could support them, even those that are decidedly unfriendly to life. So-called extremophiles tolerate extremely high or extremely low temperatures, pressures, radiations, and acidity, conditions that are lethal for other species. They invade and colonize such hostile niches as active volcanos, deserts, and the deep sea. The evidence speaks clearly: the evolution of life is not toward stability and fitness, but toward the integration of ever more, and ever more varied, elements into coherent and complex systems.

The evolution and persistence of systems in nature is rooted in cooperation rather than in competition. Competition results in advantaging some species and populations over others—the law of the jungle where everyone competes and only the winners survive. By contrast, evolution relies less on competition than on cooperation; it optimizes behavior that harmonizes goals and strategies. It does not conduce toward fitness through competition, but toward coherence through cooperation.

Direction in the Evolution of Consciousness

Natural systems are not the only kinds of things that arise and persist in the universe. Consciousness, we noted, is likewise present, and it is likewise a real and fundamental presence. We should ask whether and also how consciousness would evolve in the universe, and whether that evolution would be connected with the evolution of natural systems.

It has become clear that the answer to these questions is affirmative. Consciousness appears in biological systems already on the cellular level. Lipton pointed out that the cells that make up living systems include two classes of proteins: receptors, which are the sensory organs of the cell, and effectors, the proteins that control cellular function. The receptors "read" environmental signals, and the effectors transfer the signals into physical events; these are the cell's response. A receptor-effector complex forms a "unit of perception"—a unit of consciousness.

Evolution optimizes the capacity of cells to "perceive" their environment and to translate it into physical responses. If the evolution of this basic form of consciousness runs into constraints, cells shift from maximizing their own perception to associating at a higher level of complexity. *The Tree of Life*, an image created by August Weismann in the late nineteenth century, is a tree of growing complexity as well as of evolving consciousness.

The Evolution of Consciousness in Association with the Body

Consciousness evolves in nature throughout the range of evolution. It extends from rudimentary forms of cellular awareness to the articulate forms of consciousness exhibited by human beings. The simplest kinds of awareness are forms of irritability, displayed in the tropisms of cells and unicellular organisms. These grow in complexity, range, and articulation throughout the Tree of Life.

The overall direction of this evolution is evident in the contrast between its most rudimentary and its most evolved form. On this planet the hallmark of the rudimentary forms of consciousness is the awareness displayed by cells as irritability in regard to some features of their milieu. The hallmark of the evolved forms of consciousness is the mind-set that emerges in ethical, insightful, and spiritual human beings.

The consciousness exhibited by spiritual leaders and great philosophers, scientists, and artists is strikingly similar. Its key features are empathy, compassion, selflessness, and unconditional love. Tom Freke, a philosopher who is an authority on spirituality and a deeply spiritual person himself, noted that the spiritual traditions describe "awakening" (or "enlightenment") as an experience of all-consuming oneness and all-embracing love. His own experience testifies that this is true. In the deep awake state, he wrote, "I find myself feeling profoundly connected with all that is, and overflowing with unconditional compassion which I call deep love."*

Further hallmarks of an evolved consciousness have come to light. An in-depth inquiry coheaded by biologist Humberto Maturana on behalf of the Brahma Kumaris World Spiritual University in the year 2009 has shown that people with highly evolved consciousness possess a quality of mind that is stable, peaceful, and compassionate. Such people are dedicated to improving the life and mind of others without self-interest; they sense that their work has been assigned by a higher source and they are only an instrument; their beliefs and behaviors are consistent and integrated; they draw on a deep well of energy that gives them endless endurance and unlimited patience; they have an elevated vision of the people they serve, seeing their capacity for renewal, recovery, and progress; they have convictions so strong that they are not fazed by limitations whether in financial or intellectual support; and they have a quality of lightness, remaining available even when faced with immense tasks and responsibilities.†

In the Western world people with this kind of consciousness are regarded as spiritual leaders, and perhaps even as saints, as Saint Francis of Assisi and Mother Teresa. In the Eastern world individuals with an evolved consciousness are revered as boddhisatvas or ascended masters. They are highly respected in the indigenous world as well, seen as magicians, shamans, or medicine men (and women).

The hallmarks of evolved consciousness are present in the mind-set of leading scientists. Einstein remarked that our separateness from the world is a kind of optical illusion; Carl Jung wrote that consciousness is part of the *unus mundus,* the universe's single generative and creative

---

*Tom Freke, *Soul Story: Evolution and the Purpose of Life* (London: Watkins, 2017).
†Judy Rodgers and Gayatri Naraine, *Something beyond Greatness* (Deerfield Beach, Fla.: Health Communications, 2009).

principle, and animal intelligence researcher Jane Goodall said that she has learned that nature and herself are one consciousness. Erwin Schrödinger was explicit in defining oneness in consciousness. He said, "To divide or multiply consciousness is something meaningless. In all-the-world there is no kind of framework within which we can find consciousness in the plural; this is simply something we construct because of the spatio-temporal plurality of individuals, but it is a false construction. . . . In truth, there is only one mind."

The Evolution of Consciousness beyond the Body

Is consciousness limited to the body? If it is, it cannot survive the body. But if it is not, it can persist beyond the body, and could also evolve beyond the body.

Thinkers both East and West noted the evolution of consciousness and speculated on its nature. The nineteenth-century German philosopher Hegel viewed mind, spirit, and consciousness as a unity that evolves from primitive sense perception through religion, art, and philosophy toward absolute knowledge. Across the generations, he wrote, humans are vehicles for the self-realization of a divine spirit. Bergson viewed consciousness as the expression of an evolutionary force that strives to enlarge the potentials of the organism to choose between more alternatives, opening the way for consciousness to pass freely and to flourish. Humanity is the highest expression of this evolutionary movement, and the movement itself is the *raison d'etre* of life on this planet.

Teilhard de Chardin agreed with Bergson that consciousness strives toward ever greater freedom of expression, and also agreed that a divine spirit drives its evolution. Consciousness is pulled from the future rather than pushed by the past. Through the action of love, an evolved consciousness fuses the elements of the mind into a higher unity. Attaining this "spiritual evolution" is the meaning of existence.

A related concept appears in the "integral yoga" of Indian philosopher Sri Aurobindo. Today's human being is but a step in the continued evolution of consciousness toward the highest form Aurobindo called "the supramental." The Swiss thinker Jean Gebser described an entire complex architecture of higher and higher forms of consciousness, inspired by what he called "the Origin"—a mystical force intrinsic to all forms of existence.

Western thinkers envisaged the evolution of consciousness mainly in association with the body. In the East, wider perspectives have been adopted. Consciousness associated with the body is but one phase in the existence of consciousness, and consciousness evolves in all phases of its existence.

The above assumption is not without foundation in human experience. As we have seen, there are robust reports of consciousness beyond the brain, and even after death. Some of the reports describe the continued evolution of consciousness beyond the life of the body. They speak of the ascent of discarnate consciousness through transcendental planes, a journey that, according to Buddhists, proceeds either toward rebirth on Earth or union with the cosmic intelligence in nirvana.

A particularly noteworthy report of the evolution of consciousness in the between-life or after-life phases was furnished by the famous trance medium Rosemary Brown. It recounts the after-death experience of Frederic Myers, who was a prominent researcher of psychic phenomena in his life.*

Myers claims, according to Rosemary Brown, that from Hades, the first way station of the journey of the discarnate consciousness Myers calls "soul" ascends to the Plane of Illusion. Resting for a while in the Lotus Flower paradise, it passes to the next station: the Mental Plane. An advanced soul overcomes the longing for physical existence and strives to ascend the ladder of evolution. With the exception of those who aspire to a great intellectual feat on Earth or want to play a major role in the strife of earthly life, the soul is released from Illusion land, from "that nursery in which they merely lived in the old fantasy of

---

*Geraldine Cummins, *Beyond Human Personality* (London: Psychic Press, 1935).

Earth," and ascends beyond the Mental Plane to the Plane of Eidos. Eidos is a loftier world, but beyond is still the Plane of Flame.

Myers speaks of an ascent still further: beyond the Plane of Flame, to the Plane of Light, where reason reigns supreme. Thereafter the soul ascends to the seventh Plane, which can only be described as "Out Yonder—Timelessness." The soul dwells there not only outside of time, but outside the universe.

Our body and all natural systems evolve incessantly if periodically; our consciousness evolves periodically in association with our body and incessantly in association with as well as beyond our body, and all this evolution is nonrandom and directional. Ours is an in-formed, purposively evolving universe, and with our body and consciousness, we are an intrinsic part of it.

3

New Answers

Einstein remarked that there are two ways to live one's life: as if everything is a miracle or as if nothing is. For the outdated and but in many quarters still upheld materialist paradigm, the world is a giant mechanism functioning in accordance with laws that are in part already known and in part not yet known but are knowable in principle. When we discover these laws, the world is no longer a miracle, and nothing is a miracle in the world.

The new concept emerging at the frontiers of science offers a different view. The world is interconnected and whole, and it is in-formed by a cosmic intelligence. The systems that are present in it are evolving in a definite and discernable direction, and their nonrandom evolution suggests the presence of purpose in the universe.

This is not a finite, mechanistic-material world; it is a consciousness-infused goal-directed seamlessly whole realm. A miraculous world.

Who are we, denizens of this miraculous world, and why are we here?

Who We Are

We are conscious beings in an intelligence-infused universe. We emerged in the embracing oneness of an instantly interacting Akashic field. Our body is a cluster of vibration in that field, and it is mortal; it came into being with the fertilization of the ovum of our biological mother by the sperm of our biological father and will leave this realm at the end of our biological life. But the low-frequency, long-wavelength cluster of vibration that codes what the spiritual traditions call "soul" persists when the high-frequency cluster of our body vanishes, and it remains accessible for reassociation with another high-frequency cluster in the processes we know as rebirth and reincarnation. Our body is part of the evolution of natural systems; it is mortal. But our consciousness accesses, decodes, and transmits the infinite intelligence of the cosmos, and it is immortal. Through us, the beyond-spacetime intelligence of the cosmos enters the spacetime domain of the universe.

When our brain and body no longer function, low-frequency vibrations may still be received in association with another cluster of vibrations: another brain and body. This states and supports the age-old belief in the survival of consciousness. It completes the answer we can now give to the question who we are.

We are clusters of vibration in-formed by the intelligence of the cosmos. We are harmonies created by the plucking of a cosmic harp, ascending toward more and more encompassing melodies. The harmony of our body is transient, but that of our consciousness may be infinite. We are an infinite consciousness associated with a finite body.

Why We Are Here

We now turn to the question of *Telos*: the goal or purpose of existence. The answer we can now give is empirically based: it is grounded in the observed direction of evolution. Let us summarize the principal findings.

There are two processes of evolution unfolding in the universe: the evolution of natural systems and the evolution of consciousness.

We observe the evolution of natural systems, and we experience the evolution of consciousness. Natural systems evolve toward intrinsic as well as extrinsic coherence, and they acquire complexity in the process. The evolution of consciousness, in turn, is oriented toward the recognition of embracing oneness among the systems and the consciousness associated with the systems.

These are simultaneously unfolding processes, and they are not separate; they occur in the same spacetime domain: the universe. The evident purpose of evolution in this domain is to achieve coherence in the domain of natural systems, and embracing oneness in the sphere of consciousness.

The twin objectives of evolution are mutually reinforcing. A coherent brain and body receives and transmits the intelligence of the cosmos with more articulation and greater detail than a simpler, less coherent brain and body. In turn, an evolved consciousness provides guidance for the brain and the body in achieving internal wholeness and harmony with the external world.

Proximal Purposes

The proximal purposes of existence apply to the current condition of humanity on this planet. They can be formulated as follows.

The proximal purposes of human existence are twofold. On the one hand, they are to achieve what is known as "supercoherence" for our brain and body: the optimal coherence of the cells, organs, and organ systems that make up our brain and body. This gives us health and well-being. The proximal purposes of human existence also include achieving super-coherence with the world around us, which is optimal social and ecological alignment and harmony.

The proximal purposes of our existence encompass the evolution of our consciousness. This is an evolution that enables us to feel empathy, oneness, and unconditional love for people and for the world. It provides us with the orientation we need to progress toward supercoherence in brain and body.

Telos: The Ultimate Purpose

The evolution of our body and of our consciousness are mutually enabling: each facilitates the evolution of the other. These proximal purposes of our existence are to enable us to reach toward the ultimate purpose of our existence. That purpose—the cosmic goal or *Telos*— is to achieve the full evolution of our brain and body as well as of our consciousness.

The evolution of our brain and body and the evolution of our consciousness are coded into the very foundations of the universe. The intelligence of the cosmos acts on, "in-forms," our universe as a beyond-spacetime hologram. It in-forms the clusters of vibration that are the matter-like entities of the universe through the laws of nature, and it in-forms the clusters that appear as the universe's mind-like entities through the consciousness associated with the complex clusters we call living beings.

We human beings, as other natural systems, receive and decode the cosmic hologram. In *the internal perspective* we are elements in the flow of sensations that constitutes our consciousness, while in the *perspective of the external observer* we are in-formed clusters of vibration. Through these complementary even if qualitatively different perspectives we receive and manifest the holographically coded intelligence of the cosmos. And by acting on this manifestation, we transmit that intelligence into the universe.

Transmitting the intelligence of the cosmos into the universe is a comprehensible and meaningful endeavor. It is to enhance and activate in the universe the cosmic *Telos* of evolving toward supercoherence in body, and oneness and love in consciousness.

To integrate, harmonize, and unify all things, and at the same time embrace all things in oneness and love, is the *Telos* of all existence. It is the ultimate *Telos* of human existence. *When all is said and done, it is why we are here.*

PART TWO

The Meaning of the New Answers for Our Life and Times

4
The New Answers Re-Enchant Our Worldview

Kingsley L. Dennis

When we look up into the night sky and see the sparkle of stars, we are awed and enchanted. There is grace, there is wonder, and there is the excitement of the unknown. Everything comes alive with possibility. There is an enchanted world out there, and it beckons to us through a communal mystery. And we wish to respond to that call. For underlying all life is the urge for meaning. As human beings we desire, long for—*need*—a sense of meaning and purpose in our lives. An enchanted universe serves to entice us with a feeling of belonging. Yet somewhere along the way we lost the sense of communion.

Once, humanity experienced that sense of communion with its environment—both terrestrial and cosmic—and this encouraged a mode of direct participation. This merger between being and environment established a psychic wholeness in the human. Our ancient ancestors were not estranged from their sense of reality in the way that modern humanity has become. Human consciousness over the centuries has been undergoing a "discoupling" from the world around it, a distancing that has been referred to as *disgodding* from nature.* In short, humanity succeeded in taking itself out of the picture by creating a new and different picture of itself.

In the last few centuries especially, humankind has increasingly expunged itself out of its own mystery and thrown itself out of the realm of enchantment. The scientific, rational consciousness that has dominated our frame of reference for so long is an alienated consciousness. It views the world as if it is a separate observer, seeing a universe of objects that move in mechanical motion. This alienated consciousness has substituted the enchantment and mystery with a smear of artificiality. The cosmos of human "being and belonging" thus became infected with imposed and conditioned patterns of thinking that were restrictive and limited. For centuries we have been subjected to a reductionist, mechanical model of the universe that has dominated how things *seem to us*. Until now we have been as if in a form of mental quarantine that served to disenchant the human being from a living cosmos.

However, all this is going to change as we find ourselves on the cusp of an incredible revolution—a revolution in human thinking, understanding, and ultimately of human perception. And this is precisely what Ervin Laszlo is both discussing and spearheading in this book and in his latest writings. As Laszlo says, it is in such times of crisis and transformation, such as now, that more than ever we need to have a need to know, to remember, and to act on this new knowledge and understanding. During such times of change, the impulse for meaning and significance becomes more necessary. In such moments of social-cultural transformation, when bases of knowledge are revised, and our constructions of reality queried, the need to seek the fundamental questions of who we are and why we are here grows stronger.

The mainstream scientific paradigm, like the religious paradigm of the seventeenth century, now finds itself unable to be maintained.

---

*A phrase from Friedrich Schiller.

This is how things unfold; one set of structures, systems, and viewpoints eventually becomes outmoded and, through necessity (among other factors), gets replaced, or rather updated, by a new set. This new set then defines the dominant consciousness for the new era. New values also come to the fore to represent the new expression of consciousness. In such transitional times there is urgency, opportunity, and an interior push to reconnect with a sense of meaning and significance in life, both at an individual and at a universal level. In other words, there is a fundamental need to understand one's self and its place in the larger scheme of things. The instability we currently encounter in the world around us only convinces us further of the need to find the roots that connect us with a more permanent stream of knowledge and meaning. And this is where Laszlo positions himself, at the forefront asking whether "who I really am is what my consciousness really is." And if the human body-mind is a vehicle for consciousness rather than the producer, as Laszlo clearly shows, then we at last start to have a meaningful proposition. In other words, we begin to tackle the underlying belief that questions whether our conscious self is but a by-product of our brain.

The philosopher Gottfried Wilhelm Leibniz noted that the nature of our consciousness reflects how we view the world. Leibniz referred to the term *philosophia perennis* in his own writings in recognition of the knowledge that reflected the fundamental reality underlying our existence. The questions addressed in this volume are part of this perennial philosophy that seeks a truer understanding of the cosmos and of our place as human beings within it.

What Is the World—What Is My Mind?

The research put forward by Laszlo explains that the world we inhabit is not furnished by bits of matter. The older Newtonian paradigm has run its course. The new understanding is that, as Laszlo says, the things that furnish the world are interconnected and ultimately one . . . they are clusters of "in-formed" vibrations. Space is not the wide open emptiness that our human imaginations and popular cultural images have led us to believe. It is a dynamic interconnected field filled with fluctuations and vibrations. It is only that these fluxes of vibrations often appear to us in the form of matter. The new scientific paradigm expressed in this volume posits an underlying vibratory matrix that in-forms the material reality that our senses are familiar with. This corresponds with other insights such as the implicate order of David Bohm, the universal quantum field of quantum mechanics, and with what Laszlo calls the Akashic field. In calling it the Akashic field, Laszlo is acknowledging the debt we owe to earlier cosmologies and creation myths—an acknowledgment that gives a nod of recognition to just how right they were.

The Hindu spiritual traditions of India speak of an underlying reality that is Brahman—a reality that is eternal and eternally unchanging—where the physical world is the playground of unceasing creative play. Western streams of philosophic thought also speak of a domain beyond space and time. According to Plato there was a realm of forms and ideas—a domain of Pure Forms—beyond our space and time, and our material world is only an image or copy of this real, pure world beyond. This *pure realm* was also spoken of by other Hellenic philosophers: Pythagoras referred to it as *Kosmos,* and Plotinus as *The One.* Similarly, the Chinese sage Lao-tzu spoke of all things originating in the Tao/Dao as the unseen root of all material things. The Tao is both the originator (the source) of all things and the destination to which all things eventually return. It is the unobservable and nameless no-space, no-time, no-form essence that our words fail.

And yet this no-space is not without meaning or significance for it is infused and in-formed with a conscious, cosmic intelligence. The new scientific paradigm extrapolated from the findings presented in this book reveals that consciousness is not confined to the human brain. That is, consciousness is not a by-product of neuronal brain activity. Rather than being a product, it is a gift—a quality of vibration that is decoded by the brain from a myriad of sensory data received externally.

The human brain (mind-body) is able to both receive and transmit these vibrations that constitute consciousness. Our level of awareness corresponds to the degree to which we are able to receive, decode, and transmit the clusters of vibrations in the cosmic field. The clearer the channel to receive and decode, so to speak, the clearer is the perception—the picture—we gain. In a crude sense, it is similar to how a receiver is able to tune in to the broadcast. To use an analogy, when we first developed the television set they were small black-and-white screens, often with a fuzzy picture that needed constant retuning. Our perception of the broadcast was thus small, grainy, and without color. As our technologies advanced, we were able to produce color televisions with larger screens. As a result, our picture of the broadcast "out there" became more colorful, larger, and more involved. Now that we have high-definition, fifty-inch-plus widescreen televisions, for example, our sense of the picture is unprecedented—to the degree that we often feel we are participating in that which we are viewing.

The sense of participation in the reality around us corresponds to the level of frequencies that the brain is able to tune in to. As Laszlo notes, we can tune the brain to enter these states purposively, but we can also access these frequencies spontaneously. And the art of training the brain (or entraining it) to access and decode ever finer frequencies has been part of the wisdom teachings handed down for centuries. And yet access and decoding is only one part of this new understanding that is emerging. As a new perception set unfolds we begin to grasp aspects of the bigger picture, which is that in our known universe there is a directional impulse toward the evolution of mind/consciousness. Furthermore, since consciousness is not only received and decoded by the brain but also *transmitted,* the implications of this are that we can also consciously participate within the interconnected, vibratory consciousness field.

Until recently the concept of evolution within mainstream science was largely limited to generational biological inheritance. The understanding and the evidence for the evolution of consciousness was not forthcoming, owing to the fact that many scientists were not looking in that direction. This is similar to an old Eastern folktale featuring the inimitable Mulla Nasrudin:

> *A man is walking home late one night when he sees an anxious Mulla Nasrudin down on all fours, crawling on his hands and knees on the road, searching frantically under a streetlight for something on the ground.*
>
> *"Mulla, what have you lost?" the passerby asks.*
>
> *"I am searching for the key to my house," Nasrudin says worriedly.*
>
> *"I'll help you look," the man says and joins Mulla Nasrudin in the search.*
>
> *Soon both men are down on their knees under the streetlight, looking for the lost key.*
>
> *After some time, the man asks Nasrudin, "Tell me, Mulla, do you remember where exactly did you drop the key?"*
>
> *Nasrudin waves his arm back toward the darkness and says, "Over there, in my house. I lost the key inside my house . . . "*
>
> *Shocked and exasperated, the passerby jumps up and shouts at Mulla Nasrudin, "Then why are you searching for the key out here in the street?"*
>
> *"Because there is more light here than inside my house and so it's easier to see," Mulla Nasrudin answers nonchalantly.*

Like the character of Mulla Nasrudin in this tale, mainstream science may be guilty of spending too much time looking for the key to reality where there is more light, rather than the darkened Akashic no-time where the key actually lies. Now that we have ascertained the question, in which direction the key lies, the next question to ask is in which direction the evolution of consciousness strives.

According to Ervin Laszlo, the evolution of consciousness is a directional process, oriented toward a definable state. And that definable state is one of maximally evolved consciousness. The path toward such a maximally achieved state of consciousness is an aspect of the evolution toward ever-greater coherence. Laszlo outlines very clearly how all systems within the known universe strive toward supercoherence. The universe is quite literally coherent beyond our wildest imaginings. I don't need to repeat here the cosmological findings already presented, which show how the parameters of the universe are finely tuned beyond any possibility of randomness. There is nothing else to be said except that the matter-reality construct that is measurable shows an extreme proclivity for coherence. From the perspective of the latest scientific findings, it can be said that living forms in this universe are clusters of vibration in a universal field that is in-formed by a conscious, cosmic intelligence. While this may represent the fundamentals of our world, it is only the beginning in regard to who we are and our purpose. According to Laszlo, this purpose is to *transmit higher consciousness into the universe.* And this compels us to embrace the values and qualities of understanding, empathy, compassion, unconditional love, and unity. And it is these values and qualities that lend meaning to our existence as human beings. To strive toward attaining these qualities is part of the coevolving process that we each participate in between ourselves as human beings as well as within the larger cosmic order. And in this striving we grow in our understanding of who we are and why we are here.

Who We Are—And Why We Are Here

To leave the dead dinosaur at rest, let it be stated clearly here that life on this planet is not an accidental, random anomaly that emerged on a rock hurtling through dead space. Although in more recent times some scientists have come to the conclusion that there is an order to the universe—a form of intelligence—they fall short in positing a purpose. That is, they stay clear of stating that evolution is a directional process toward any definable, purposeful goal. This book declares otherwise, and posits a different outcome acquired from the latest in scientific research. Rather, that the intelligence behind the known universe informs its evolutionary processes with coherence, consciousness, and unity. It is not a new worldview so much as a new view of the cosmos—a new *cosmoview,* if you will. And this new perspective views existence not through the duality of "other-object" but through a receptive mode that sees all reality as inherently interconnected and interrelated. This view is similar to that which we possess, as babies, when we first arrive into the world.

According to many child psychologists, such as Jean Piaget and Arnold Gesell, children are not born into the world with an *object self.* It is only around three years of age that children begin to recognize and be interested in other children. That is, they begin to recognize the distinction between self and other. The receptive self then shifts into the objective self, and the world expands enormously for them. Yet at birth the receptive mode is dominant, but through gradual maturity it is replaced by the objective mode that guarantees survival. We need the objective mode in order to navigate the obstacles we encounter in our lives, and in order to successfully deal with the challenges of a physical existence. Yet this useful survival mode has, over time, taken the role of being the dominant feature of the human self. In large part this is because such a behavioral mode was further reinforced by a materialistic, scientific (and later social) framework that emphasized duality and mechanism. We should not forget, however, that the receptive mode is an important part of the psyche of the human being. It enhances our capacity for communication, interrelation, collaboration, understanding, caring, and empathy. It helps to keep a healthy balance between self and service, and forms a coherent perspective between that which is solid and those things more ephemeral and sensual. The receptive mode emphasizes the *C* values of Connection ~ Communication ~ Consciousness ~ Compassion.

The receptive mode is more prominent in acts of selfless service, yet such perspective and understanding may not be a customary one, especially in Western societies. An old tale reflects this:

His Own Suffering

Whenever the rabbi of Sasov saw anyone's suffering, either of spirit or of body, he shared it so earnestly that the other's suffering became his own. Once someone expressed his astonishment at this capacity to share in another's troubles. "What do you mean 'share'?" said the rabbi. "It is my own sorrow; how can I help but suffer it?"*

Contrary to this, the object mode emphasizes the alternative *C* values of Competition ~ Conflict ~ Control ~ Censorship. Because the object mode/object self has had the dominant role in our recent history, these alternative *C* values have become the norm. They have subsequently been reinforced and adopted into modern materialistic lifestyles so that now each complements and reinforces the other. To step outside of this worldview has previously meant ridicule, harassment, and even possibly death. However, the tide is now turning, and this focus on control and competition is at variance with the new paradigm understanding of an interconnected, unified reality. The incumbent paradigm is now defunct and can no longer serve as the default form of perception and cognition. It is for this reason that we are likely to see in the years ahead great change sweeping through our diverse human societies.

It is my own view that the twenty-first century will witness the equivalent of the dramatic "flat Earth to round Earth" shift that transformed our human understanding—and our societies—centuries before. Some commentators refer to our current time as witnessing a Third Industrial Revolution; others say it is a Fourth Industrial Revolution.[†] Yet rather than referring to this transition as an industrial one, I consider this profound shift as a Revolution in Human Being—or rather as a *Revolution in Human Becoming.* The possibility of a genuine planetary civilization, with unity through diversity, was never in the cards—until now, that is. I propose that we have entered a phase where there

---

*Martin Buber, *Tales of the Hasidim: Early Masters* (New York: Schocken Books, 1948), 86.
†See Klaus Schwab, *The Fourth Industrial Revolution* (New York: Crown Publishing, 2016).

will be new forms, new arrangements, new structures, new perspectives, and new emerging states of *being*.

The understanding provided by the new answers will serve to re-enchant our worldview. As such it will assist in the birthing of a new era with its attendant values and qualities of connection, communication, consciousness, and compassion. However, at this present time those of us caught up in the disruptive wave of changes sweeping through our societies need to remain grounded, balanced, and work with the disruptive changes rather than against them. Change on this planet will come through us, the people, and the attitudes, awareness, compassion, and sincerity that we embody and manifest. This is the real stability that can be passed on to those around us: our family, friends, communities, social networks, and so on. It is imperative that those of us working toward betterment and change maintain a focus on the great opportunities emerging now and not become disheartened by disruptive forces. People's minds are changing all over the world, and with this can come remarkable and inspiring potential and opportunity. Things can no longer go on as they have been; this is now plainly obvious to most observant people.

The new paradigm of reality represents a stage in which the energies of coherence and harmony will be recognized as being principle drivers behind human, planetary, and cosmic evolution. The understanding that is unveiled through the answer given in this book will ultimately bring renewed meaning to the question of who we are and why we are here. We are each of us striving for self-significance, which in the end will come through greater participation with a reality that is greater than each of us and yet contains us all. To participate effectively—*to transmit higher consciousness into the universe*—we require the qualities of empathy, understanding, compassion, unconditional love, and above all, service. Through service to our fellow human beings and to our societies we are simultaneously serving the drive toward maximum coherence. The beginning step on this path is to accept, adopt, and utilize a coherent view of reality.

The Implications of the
New Paradigm Understanding

By the time we look back on this age from a new phase of human civilization, historians will see the Cartesian-scientific paradigm as a relic. It will be viewed as a curiosity of mind that created rapid industrial expansion and scientific knowledge, yet failed to bring real progress within the essence of the human being. Its spurt that lasted several centuries may be likened to a primitive rocket booster that propels a spaceship into orbit only to be spent and cast off, to fry up and dissolve as it falls back down to Earth. The last few centuries were a single evolutionary episode that ran its course. In anthropological terms, it was a mere blink of an eye. And in that blink, humanity brought itself to the brink of collapse. Yet, at the very last step, it will be seen that humanity pulled itself back on the path, as a new evolutionary epoch pushed its way through with disruptive labor into planetary birth. We are in the midst of that birthing transition now. We have the discoveries of new science mingling with new technologies. The digital/virtual worlds are augmenting our sense of material reality, and the deep cosmos is exploding into revelation and being revealed. The great sacred mirror of the human self is reflecting back to us every known atom that ever sprang out of the creative matrix of existence. It is the age of momentum, acceleration, exposure, disclosure, invention, innovation, exploration, and self-understanding as never before seen on such a widespread scale. The new map of reality will assist in this momentum to give us meaning and to empower the individual within a grander unified whole. Where once a higher state of consciousness was first developed through exceptional individuals, it is now on course to be developed through the masses by receptive individuals and communities connecting together across physical and temporal boundaries.

The implications of the new paradigm are that we come to realize the known cosmos acts as a whole nonlocal, unified consciousness field, of which sentient life-forms are localized manifestations. It has been inferred through various religious and sacred texts and traditions that the universe (material reality) came into being as a way for its Source "to know itself'"; "*I was a hidden treasure and wanted to be known.*" This is reminiscent of "*Know thyself,*" the famous maxim from the Delphi oracle. Or, as in the words of philosopher Henryk Skolimowski, "We are the eyes through which the universe contemplates itself. . . . We are cosmologizing the human."* Self-consciousness is ascribed to those sentient beings at the peak, or greatest actualization, of mental development. Self-realization is something we credit to each attained individual consciousness. A realization of the self is part of the path of human actualization; it is a path in which purpose and meaning are core drivers and potentials to be achieved. It is an inner knowing that defies the orthodox scientific view of "selfish genes." Human beings—as human *becomings*—are naturally driven by a longing, a purpose. A part of that purpose can come through the knowing that we, as manifestations of human consciousness, are localized expressions of the greater unified cosmic consciousness.

As sentient beings we receive aspects of the cosmic consciousness that pervades our spacetime—we are *animated by it*—and we then manifest this through our own socialized minds and cultures. Our individual expressions of consciousness in our reality matrix also reflect back into the unified cosmic consciousness, thus serving to enhance the consciousness of the cosmos beyond space and time. The sacredness of being human is that we each have a role in bringing the unfinished material reality into greater coherence, and thus completion. If enough localized consciousnesses awake on this planet we can catalyze a localized planetary field into conscious awareness. That is, a planetary matrix field is sufficiently prepared to receive, "bring in," the greater consciousness pervasive in the cosmos (comparable to the "immanence of the Supermind," in Aurobindo's terminology). In this case, we are

---

*Henryk Skolimowski, *The Participatory Mind: A New Theory of Knowledge and of the Universe* (London: Penguin/Arkana, 1994), 3.

each a conscious agent of cosmic realization and immanence. We each have an obligation in our existence on this planet to raise our individual, localized expressions of consciousness. In doing so, we both infect and inspire others in our lives to raise theirs, as well as reflecting back our conscious contribution into the source That Is. In this way, we can act as both citizens of the cosmos as well as caretakers for the sacred cosmic order.

This sacred cosmic order informs us that our reality is not a static state but an active, fluid realm that makes demands on us. We are on a path of completion—of conscious completion and communion—that is the eternal path of the sacred. Through this sacred journey of completion we connect and commune with everything else in our reality matrix—and *beyond*. We can achieve this through our small acts of conscious participation, and regain our communion with the cosmos. The emerging technologies and social change on this planet may well be part of this process, in-forming an extended mind and empathic embrace across the face of the Earth. Everything is ultimately a technology of the soul—and all understanding, all science, and all human expression is a tool for moving closer to soulful communion with a grand conscious and sacred order.

The new answers put forth in this book give us hope that one day we may witness a grand awakening, unprecedented on this planet—and this may very well be the purpose for sentient life, as conscious agents of the sacred cosmic order. This is likely to be more reality than fantasy. The hidden treasure that is at the very core of our existence wishes to be known—for *us* to be known—by our individual journeys of self-realization.

To summarize, the new paradigm understanding will catalyze new value sets, systems, and institutions on this planet—in alignment with the drive toward coherent, conscious evolution. Our responsibility now is to fully engage and be a part of the human *becoming* that we truly wish to see in the world. This requires that we spearhead the transition at hand and that we show, through our behavior, the new models for change. We have the opportunity, and capacity, to do this for ourselves and, more importantly, for those to come—and this shall be our true legacy.

> *As proof that we are regenerated, we must regenerate all around us.*
>
> LOUIS CLAUDE DE SAINT-MARTIN

5

The New Answers and the Power of Purpose

Emanuel Kuntzelman

We are living in the most critical time in human history. And there are two main reasons for this grand affirmation: firstly, we live in an age in which science has given us the essential understanding of who we are and why we are here; secondly, our civilization is now facing the greatest challenges we have ever seen. The explanations for the first reason are eloquently expressed by Ervin Laszlo and the contributing authors of this work. Sophisticated technologies have allowed us to explore the microscopic makings of the quantum world, revealing a fascinating realm of the in-formed Akashic field underlying all reality. As for this being the most challenging time in history, one only needs to look at the exponential growth of population and environmental deterioration combined with the uncontrolled increase of armaments and material greed to conclude that, unfortunately, there has been no other time that remotely compares with today.

The new paradigm of science that reveals interconnected oneness and instantaneous, nonlocal communication confirms the positive side of Einstein's proposition that life is, indeed, miraculous. The universal mind of consciousness is infusing existence with meaning every step of the way. On the other hand, we could say there is a five-hundred-pound gorilla in the room that very few of us recognize. Whereas Jane Goodall informs us that gorillas and chimpanzees demonstrate a sense of awe at the wonder of nature, inclined as they are to marvel at the beauty and mystery of a waterfall, we humans seem to be so entranced in our material world that we have lost the ability to appreciate our natural surroundings. So many of us no longer wonder why; we simply focus on getting more material wealth before the Earth's resources are exhausted. The "hard problem" then is not so much who we are. The evidence for an intelligent field of consciousness is becoming so extensive that even materialist science will soon have to acquiesce and admit its existence. The harder problem seems to be why are we here? Laszlo gives us a short and definitive answer to aspire to: we are here to transmit more and higher consciousness into the universe. Unfortunately, for many of us the meaning of our existence goes no further than a desire to accumulate material possessions at our neighbor's expense.

This book postulates that the very fabric of the universe is a purposeful field and that we are the products—perhaps the crown jewels—of this purposeful evolutionary progression over the last fourteen billion years. If that is the case, we are a long way from even understanding that meaning, let alone bringing it to fruition. Goodall refers to Laszlo's observation that most people are evolved enough to know that they need to change, but not evolved enough to know what change they need. Thus, the hardest problem of all might be, as Laszlo states, that many people, including even scientists, do not see what they do not believe. Let's address this lack of belief in order to better understand if there is something new in the field that we are simply not seeing.

Big problems, such as the ones our civilization faces, need radical solutions. The radical postulation in this case is that the failure to believe in ourselves stems from the limitations of our cultural conditioning. Realizing that most people's attachment to their own culture borders on the sacred, we are going to go out on a limb and propose that it is now time to, if not slay that sacred cow, at least recognize it in order to transcend to a higher level of cultural evolution as a participating species that shares our planet with the rest of the inhabitants of this marvelous ecosystem.

Culture is near and dear to us because it has given us meaning, beauty, and even survival. In the early days of our existence, the culture of clans brought us together in small communities so we could better protect ourselves from the dangers of the environment. Within the emerging human social groups, art forms slowly developed, from cave drawings to symphonies. Language, music, literature grew from the cultural petri dishes of small collectives who cautiously joined others to form richly diverse cultures on our planet. After millennia of painstaking development, it is no wonder that the wide array of cultures we have today takes pride in what they have accomplished. Our cultures represent the many achievements of the human will, first to merely survive, and later to symbolize the flourishing aspects of our existence.

Yet for all this significance, most of our cultural accomplishments remain at the ethnocentric level of development. Politically and administratively, we struggle to go beyond the nation-centric level. Because of the deeply emotional history of our culture, we all tend to be overly attached to our separate identities. We are often most proud of our own art, language, customs, religious beliefs, and social practices, preferring our own cultural identity above all others, without the ability to objectively evaluate cultures beyond our own. We are unable to completely see the wonders of others when the pride and prejudice of our ethnocentricity masks our eyes.

The new paradigm of scientific understanding described in these pages desperately needs to be accompanied by a leap to a more inclusive and purposeful cultural paradigm. If "living systems are astonishingly coherent," then the living system of humanity has a large need for improvement. Taking an honest look at the divided world of nation-states and bitterly opposed religions and ethnic groups that we have today, we are clearly far short of anything close to coherence in our social and cultural norms of behavior.

What we need is to upgrade our worldview from ethnocentric to world-centric—gaining a planetary perspective that accepts humanity as a whole and gives priority to the health of our Earth's ecosystem. Unquestionably we could benefit from a more robust system of world governance that would override the interests of separate nation states. A planetary administration, however, will only be possible when we begin to move away from the separatist and protectionist policies of our ethnocentric worldviews.

Because it will be a huge undertaking to move planetary culture up to a world-centric point of view, we can best serve this goal by shooting high—for the stars, so to speak. This is where the underpinnings of the new paradigm's deeper understanding of the meaning of existence play a crucial role. Up to this point in human history we have answered the question "Why are we here?" in ways that are not entirely satisfactory. On the one hand, we have scientific materialism saying that we are nothing more than thinking blobs of meat that accidentally evolved out of a long line of random mutations. That scenario negates a purposive motive for our existence and provides no support for the search for meaning in our lives. Small wonder, then, that if we assume we are nothing more than the fortuitous accumulation of material wealth as a species, the accumulation of material wealth is the main driver of our culture.

On the other hand, through our diverse religions we have arrived at cultural interpretations of a more creative nature regarding our universe, but these ideas lack the sophistication of the twenty-first-century understandings of physics and cosmology. Monotheistic traditions have provided us with a highly anthropomorphic view of a Divine Being that carries the weight of the foibles of human attributes. Eastern religions have given us a more subtle and organic view of creation, but they still do not provide a strong evolutionary concept of existence. They tend to caution us of the dangers of any hope for the eventual survival or development of an individual soul while assuming that there is a cyclical nature to existence rather than one that contains intentional directionality.

The new paradigm as described here provides the cornerstone of a more sophisticated understanding of who we are and why we are here. This approach takes the middle ground of utilizing the scientific evidence for purposive evolution and combines it with a conscious (spiritual) background of reality and gives a highly compelling answer to these great questions. We begin to understand that we are the Omega Point of a long, teleological process that developed into us in order to consciously contribute to the process itself. We are here to be pulled into the future, to integrate the trajectory of our individual pursuits with the holistic purpose of the universe, which is to create a higher consciousness.

Laszlo presents some intriguing evidence for the survival of our consciousness beyond the necessity for it to be connected to the brain and the physical body. As such, there are no barriers to the notion that our individual consciousness—our soul—continues to develop without the requirement of a body. When death brings our life to an end, it is merely the end of a material appearance to the evolving soul behind the scenes. This explanation is perhaps the most compelling contribution to answering the question of why we are here. The reason for our existence is to evolve through a series of incarnations and improve the vibrational level of the cosmic ground, the Akashic field.

The accounts and reports presented in this book about communication with the deceased are only a small portion of the evidence for life after death and for life in between a continuing series of incarnations. The documentation on cases of possible reincarnation compiled by the late Ian Stevenson, and now continued by his protégé Jim Tucker at the University of Virginia, provide convincing reading for anyone interested in knowing more about the subject. Brian L. Weiss has logged an impressive account of patients telling their stories of past lives through hypnosis and psychotherapy, and the late Michael Newton used similar techniques with hypnosis to enable his patients to explore the realm of existence between lives. The fascinating case studies detailed in his books *Journey of Souls* and *Destiny of Souls* reveal a rich world of higher consciousness beyond the physical realm, a world strikingly similar to the description provided by Frederic Myers from the other side, as well as the accounts contained in the *Tibetan Book of the Dead*.

For anyone looking for purpose in life, these extensive studies give compelling reasons to affirm the existence of an afterlife, perhaps enough to persuade the most recalcitrant of critics such as materialist scientists and the likes of Bertrand Russell. For those hoping that they could get away with frivolous behavior in life and simply be done with it all at death, the news is not so good. These studies make it apparent that every detail of our lives will be worked through in our existence between lives. There is a very strong argument that such a thing as karma does exist, that every thought, word, and action we undertake in life is recorded in the Akashic field and remains with us as part of our evolving soul or essence. All of this karmic information is stored in the vibrating wave clusters that form the reality of the cosmos.

So, the news is good for those who believe that there is meaning underlying our existence. It is so good that it could be shouted from the rooftops: everything we do, with all the intentions coded into our acts, is absorbed and stored, in-forming the cosmic field so that in our existence between lives, and in our incarnations yet to come, we can continue to evolve and perfect our contribution to the evolution of consciousness. The bad news is the same: every willful act of our consciousness becomes part of the fabric of the universe and we shall eventually have to reconcile our actions with the deep meaning that underlies our existence.

All of this, for better or worse, gives a profound sense of purpose to who we are and why we are here. It provides the basis on which a new worldview can be constructed—in this case, a cosmo-centric view. We can understand that our universe started with the intention of developing a more evolved consciousness and has developed toward that goal since the beginning of time. We are now an integral component of that grand entelechy, plugging into our *elan vital* to become the highest expression—the finest form of vibration—that we can possibly be.

And with this emerging cosmo-centric view comes a new appreciation of the highest values ever attributed to our species. Truth, beauty, and goodness can become the framework of our actions. Peace, love, and understanding return to take their place as the pinnacle of our aspirations as human beings. With these reenergized higher values at our service, we can readily bring the cosmo-centric view to front and center. If we then apply the understandings and values of this higher consciousness to the ethno- or even egocentric worldviews so prevalent today, we can surely draw ourselves upward, creating a quantum leap in the overall cultural level on planet Earth.

Rather than denying or abandoning the cultural accomplishments we have achieved over thousands of years, we can include and transcend them. With the enlightened view of identity and purpose established in the findings of new paradigm thought, we are given a new opportunity to understand our lives as one big miracle. The gateway is opened to an expanded explanation of our place in the cosmos. We are central to it. We are the evolving cosmos itself. Instead of being obsessed with our material survival and overly protective of the culture of our clan, of the identity of our nations, and the absolutes of our religious orientations, we can become true world citizens and appreciate the rising culture of the planet. It is this emergent culture that holds the health of our earthly ecosystem as the number one priority. We can only be as healthy as the home we live in. The emerging worldview culture should also take into account the lessons of the new paradigm. We are, after all, an interconnected whole, striving to work together and establish coherence. As a planetary culture we can be of one mind, finding a way to preserve our ecosystem and flourish as a whole.

As we delve ever deeper into the implications of new paradigm research, it becomes apparent that neither reductionist science nor fundamentalist religion are culturally evolved enough to provide an ethical and moral base to fulfill the needs of the great transition we currently face. In their different ways, both science and religion will eventually be able to offer a more complete sense of purpose as they integrate the knowledge that is available to us today and is so eloquently presented in this book.

In the beginning, there was purpose. That is the underlying message of this book. Before the Big Bang, there was a primordial field of consciousness. Whether it was in the form of a metaverse giving birth to our baby universe, perhaps through the umbilical cord of a black hole, or in a massive download of light emerging from the bursting potential of consciousness, or any of the other beautiful scenarios of our possible beginning, it burst forth with a sound and a fury that signified everything.

For any thought to manifest, it should have a purpose. Perhaps there might be occasions when the worst part of our monkey minds sometimes "thinks" about something without a purpose, but it seems unlikely. Even our most visceral and reactionary emotions, when we apparently react without thinking, have a purpose, albeit not a very good one. The bigger the thought, however, the more likely it has a deep, beneficial purpose. Something on the scale of creating a universe must have had a truly profound purpose. Far from being a material accident, the creation of our universe was the manifestation of purpose in its highest, and most conscious, essence.

As conscious individuals, we experience this purposeful arousal of thought thousands of times every day. Each time that we generate a new thought, it comes from the field of potentiality of our own being. In the same way, the inception of our physical world came into being as a result of producing an action in the primordial field of consciousness. The field thought, and therefore it was. And that original thought has arisen from purpose, from the intent to come into being and create a platform from which consciousness could continuously evolve.

Some years ago, I was shown a video at a personal development conference and was told to count how many successful passes were made by a group of people frantically throwing basketballs to one and another. Always up for an intellectual challenge, I focused all my attention on calculating the number of passes. What I didn't notice was the five-hundred-pound gorilla that strolled through the middle of the scene.

Nor did anyone else in the audience observe the gorilla. When we were allowed to watch the video again, it was astonishing to see how we could have missed the gorilla the first time through. There he was, calmly waltzing through the scene for a full nine seconds, and everyone was so absorbed with the task at hand that they failed to notice him.

Concentrating as we were on getting the "right" answer, none of us saw the anomaly that was the point of the exercise. Reductionist scientists, like all of us obsessed with counting passes, have experienced their own version of this exercise as they focus on measuring anything that is measurable: particles, atoms, molecules. In so doing, they are good at adding up the parts, but are not so adept at seeing the whole.

Indoctrinated into this scientific culture, human beings in general often have difficulty seeing anything other than what they already believe. We can easily count up the parts, but we have difficulty seeing the whole picture unveiling right before our eyes. So concerned are we with short-term gains that we fail to comprehend the long-term consequences of our acts. We are well aware of the balance in our bank accounts and woefully ignorant of what we are doing to future generations. This all comes partially as a result of being conditioned by the materialistic culture of the modern world.

When we are able to escape enough from the daily grind to be able to contemplate the idea of the presence of a divine spirit, we often resort to utilizing a cultural concept of a deity commonly known as "God." Even if we don't fall into the trap of visualizing this being with predominantly human male characteristics, we are impeded by these cultural underpinnings from fully comprehending the true sense of something greater, a nondual energy that originally infused the universe with directional purpose. Words usually do not come to us to adequately describe higher consciousness, whatever name we give it.

With the knowledge now available to us from research and studies regarding the quantum nature of the universe and the Akashic field, it becomes increasingly apparent, however, that there is good reason to believe that we live in a purposeful and meaningful universe. *Nonlocal communication, entanglement, the uncertainty principle,* and now *super-coherence* are among the terms commonly used at college level studies. No longer are the mysteries of the universe hidden from us in the esoteric realms of quantum physics and cosmology. Thanks to the tireless work of Ervin Laszlo and many others, we can more readily comprehend the vibrating and energetic nature of the universe and apply it to a comprehensive cosmo-centric worldview that will come to our rescue when we are bogged down in the quagmire of separate cultures.

Knowing that our purpose is to evolve and contribute ever higher consciousness allots a full sense of meaning to our existence. Recognizing that the evolutionary process is applicable to the development of our consciousness and that we likely live a multitude of lives adds even more profound purpose to who we are and why we are here. It makes us realize that we are responsible for every creative act, whether thought or deed, that we produce as living beings. As products of the Source field of consciousness (God, Origin, Zero Point Field—however we care to label it), we are cocreating the new consciousness with all of humanity and the Source itself. We are one, and we are all in this together. We therefore must welcome with widely open arms a shift to a cosmo-centric view that provides the framework for embracing the higher values that we have always held dear, but have forgotten in our efforts to count up everything measurable along the way. All of this paves the way for a worldwide cultural transformation that will open a path toward living in harmony, peace, and supercoherence.

In short, we have greatly expanded our belief system and can now see the gorilla in the room—and recognize that he is our friend. Harbingers of danger on the horizon are rarely welcomed and are not often even recognized. Now, however, in this singular moment in history, with the exponential growth of population, technology, and armaments coupled with the destruction of our environment and the dramatic loss of species, there is no doubt that there is trouble in the air. We need to reestablish a friendship with our fellow inhabitants of Earth and listen to what they have to say.

Let us assume, for example, that our gorilla is named Koko, like the now famous primate in the video that has been watched by nearly two million people on YouTube. Koko, born in a zoo and raised with humans, has learned to use the hundreds of sign language words that Jane Goodall mentions as a way to communicate with us. Koko is still in awe of nature, and metaphorically beckons to us to sit down next to her and contemplate the beauty of a waterfall. There we sit on the riverbank with her, contemplating the splendor of the scene. Scintillating golden rays of sun are filtered through the mist of the cascading water, hurling toward us in shafts of rainbow light. The splash, ebbs, whirls, and flow of the water lure us into imagining the river as the symbolic stream of time, as an evolutionary force that brought us to this miraculous moment.

Koko turns and looks with sad eyes into the heart of humanity. In her simple, gorilla sign language, she indicates that what bothers her most is that we humans have lost our connection to nature. Unlike the apes, we find it difficult to sit and marvel at the wonders of our world. With the cultural and social backdrop of the industrial revolution fogging our thinking for hundreds of years, we see nature as separate from ourselves, almost as an adversary, certainly something to be conquered. But why, wonders Koko, and those *Homo sapiens* who understand the coherent unity of things, would a species ever do anything to destroy its own environment? Why would we not cherish the generations to come and our own future lives and do everything possible to preserve our environment, keeping it clean and fresh, just as we do our own home?

I look at Koko and send my thoughts her way with sincere sympathy in my eyes. With our new appreciation of the majesty of existence, of knowing who we are and why we are here, we have a renewed conviction to cooperate and do our part to ensure the future health and well-being of our planet and all of its species. We get it, Koko. We know the time has come to elevate our cultural consciousness to a level worthy of the magnitude of this incredible experiment known as the cosmos. If the Source field could fine-tune so many factors to a billionth of a degree to make all this happen, surely we can and will do our part to make sure that the process continues, that the river of time will keep on flowing and filling us with awe.

Looking at the river at the base of the waterfall reminds me of another story. It is a tale of the simplicity of transformation. While it seems like something that is almost impossible to achieve, the basis of transformation comes with a mere change of perspective. Some five hundred years ago, the Renaissance transformed art and culture in Europe, and eventually most of the civilized world. It all began with a handful of artists who figured out how to put more perspective in their paintings. The third dimension of depth had always been with us, but they were the first to represent that additional dimension on a flat canvas, and thereby trigger a movement that would motivate society to look deeper and extrapolate more meaning from the flatland of our lives as we emerged from the Middle Ages. Today we are in a similar situation, yet with a highly amplified importance. We need a new perspective to help us understand that if we believe differently, we see things differently. If we believe in the rising mountain of facts that tell us that we live in a field of evolutionary consciousness driven by purpose, we could easily see that we have always known this and have always lived with this understanding in our hearts.

Now, back to our concerns. They have to do with Mulla Nasrudin, who also makes an appearance in this book in the contributions by James O'Dea and Kingsley Dennis. Somehow the tales of this thirteenth-century character, with his humorous simplicity and reverse logic, seem appropriate in the current times of global confusion. As it happened, a group of friends were out on a trek in the forest, headed to a new, yet unknown destination. They came upon a raging river, much like the metaphorical waterfall that we were enjoying with Koko. There seemed no way they could safely cross the waters. Then, on the opposite bank, they espied their old friend Mulla Nasrudin.

"Mulla, Mulla," one of them shouted. "How do we get to the other side?"

The mulla looked around, a little disoriented by the question, and responded, "But you are already on the other side!"

When we obtain a new visual perspective on something, it creates a shift in our brain. If we look at one of those drawings that can be seen in two different ways depending on how we look at it—there is one, for example, where we see either an ugly elderly woman or a pretty young lady—and we manage to see the other angle that we did not see at first, this produces a burst of high-frequency gamma waves in our brain. An "Aha" moment of inspiration also brings the same effect with a surge in high-frequency brain waves. Sometimes a relatively simple, yet new, insight on our situation is all we need to not only see things differently, but also feel refreshed and renewed. It is likely that at the instant of the Big Bang there was a huge burst of creativity and an outflowing of gamma waves. The grand thought that initiated our universe is very likely reiterated on a smaller scale every time we realize that the whole world changes by the simple act of looking at it all from a different point of view. This is what the new paradigm principles do for us.

Thus, we could say that the only barriers to the potential transformation of our consciousness are those of our own making. The Renaissance brought three dimensions into our lives. Einstein gave us the fourth dimension in the twentieth century. The findings in new paradigm thought, brilliantly exposed in this book, now give us an additional dimension—that of a cosmic intelligence. The light of higher consciousness shines so brightly right where we are that we only need to blink and see ourselves for what we truly are—the pinnacle of evolution toward coherence.

Margaret Mead's famous quote about a small group of like-minded people being able to, and in fact being the only way to, change the world comes to mind. Do not think that the relatively small group of people now articulating the new paradigm of purposive evolution of consciousness cannot do the same. Indeed, it is the way it will happen. When we tap into a great idea whose time has come, it becomes the attractor in the hearts of humanity and leads us out of chaos. As a rising tide lifts all boats, the emerging cosmo-centric worldview brings a new cultural understanding of the meaning of our existence and pulls us forward into the greatest transformational shift in the history of humanity.

We always knew that we are part of a purposeful project. Now we perceive that project with all of our senses. Sitting here, on the river bank with Koko, we know that we are on the right side, the only side, of change. We were here to begin with, and now, for a restful moment, we can allow ourselves the comfort of sitting here with eyes filled with tears of joy and be in awe of the miracle we know as nature.

6

The New Answers and the Goals of Contemporary Social Change

Maria Sagi

In giving an answer to the perennial questions of who we are and why we are here, Ervin Laszlo suggests that there are two basic processes of evolution in the universe: the physical and the mental. We observe the evolution of physical things—natural systems—and we experience the evolution of consciousness. The purpose of this two-pronged evolution, he tells us, can be discovered. In regard to natural systems, this purpose is to progress toward supercoherence: intrinsic coherence within the systems and extrinsic coherence between the systems and their environment. On the level of consciousness, the purpose is to evolve our individual consciousness toward the level hallmarked by responsive and responsible communication, empathy, and unconditional love.

In light of the new answers, the basic purpose and goal of contemporary social change can be defined as supercoherence in the world at large and an evolved consciousness in human beings. These are meaningful and important goals and they need to be pursued.

The Goal of Supercoherence

Supercoherence is a relatively new technical term that is not yet widely known. It is, however, highly relevant to today's world. Meeting the condition of supercoherence in society would mean having the skills, the organization, and the motivation to face and overcome the problems of today's world.

Coherence means a form of organization in a complex system where every one of its parts is effectively linked with every other part. Applied to a given system, this means internal or intrinsic coherence. Coherence between the given system and other systems in its environment is external or extrinsic coherence. Internal coherence spells viability and health in the system; external coherence indicates adaptation between the system and the world around it. Both are needed if the system is to persist and flourish.

Systems in today's complex world are "holons" to use the term suggested by novelist Arthur Koestler. Holons are Janus-faced entities: they are whole in regard to their parts, and they are parts in regard to other systems in their environment. In their "whole face" holons are intrinsically coherent: they are a coherent, mutually fine-tuned ensemble of their parts. In their "part face" holons are extrinsically coherent: they are adapted to and with their surroundings. A true holon is both intrinsically and extrinsically coherent. Its condition can be described as supercoherent.

Systems in nature have achieved supercoherence or are approaching it sufficiently to persist. A living organism must be intrinsically coherent if it is to persist in the physically improbable and intrinsically unstable ("far-from-equilibrium") state that characterizes biological systems in general. Intrinsic coherence for a biological system is a precondition of its viability. Achieving it means health for the system. A living system must be also extrinsically coherent. It is vitally dependent on systems in its surroundings for the resources it needs to live and survive, beginning with the air it breathes and the water and the food it accesses and ingests. Extrinsic coherence spells adaptation of the system to the niche in which it finds itself.

Intrinsic and extrinsic coherence together mean supercoherence, and supercoherence is a basic requirement for living systems on this planet. Not all systems achieve it, and those that do not suffer the consequences. A lack of intrinsic coherence means ill heath: breaks and blockages in the relations among the cells, organs, and organ systems of the complex biological system. A lack of extrinsic coherence in turn means a breakdown in the adaptation of the system to its surroundings. In systems in nature a failure in either of these regards leads to the demise of the individual system and ultimately to the extinction of the species. Natural selection weeds out unhealthy and poorly adapted systems and it replaces them with better adapted species. But natural selection does not work as the highest arbiter of persistence and extinction in the case of humans and human societies. Here an entire array of increasingly sophisticated technologies intervenes: they compensate as far as possible for flaws in the systems' coherence. Medical technologies and those oriented toward preserving health and well-being compensate for inadequate intrinsic coherence in human beings and societies of human beings, substituting physical or biochemical means for natural aptitudes in creating and maintaining the system's internal coherence. Technologies for the manipulation of the environment compensate in turn for a lack of natural adaption in individuals and societies. No longer are human systems constrained to adapt to their environment the way they find it: they can actively shape and reshape their environment to fit their own needs. Unfortunately this endeavor does not stop at the limits of actual needs: it embraces a growing range of culturally inculcated artificial needs—actually, wants.

Technological compensation encounters outer limits. The use of medicines and medical technologies can go only so far in preserving or reestablishing health; beyond a critical point they are useless or create adverse effects. There are corresponding limits to our ability to manipulate our food supply and access clean air and potable water. In the final count, technologies are temporary remedial instruments, not lasting and reliable solutions. There is no true substitute to organic health and a high level of adaptation to our surroundings. No substitute, that is, to our *supercoherence*.

In today's world the most visible signs of the unfulfilled need for intrinsic coherence are the spread of diseases and the malaise of large segments of the global population. The lack of extrinsic coherence, in turn, shows up both on the biological level of maladaptation to the environment and on the social, economic, and cultural levels in the relation of individual societies to their social, economic, and cultural milieu. These forms of maladaptation are growing rapidly. In the political sphere their most viral symptoms are ultranationalism, keynoted by slogans such as "my country first," recalling the midcentury Nazi slogan *Deutschland über alles* (Germany above everything). In the economic sphere the equivalent condition is indicated by the shareholder philosophy: that the sole responsibility of businesses is to make money for their owners. In the cultural domain the corresponding flaw shows up as fundamentalism. My tribe, my belief system, my race is the indicated focus of my aspirations and activities without much concern for the aspirations and well-being of the others.

The consequences of these faults in our intrinsic and external coherence become more and more evident. The time has come to reconsider our view of the world and our way of relating to the world. We are not separate, mutually detached parts in a soulless machine, but interacting and interdependent parts in a finite and interconnected living world. We need to become coherent in us and coherent with our world. That is, we need to become supercoherent. Progress toward supercoherence is the meaningful direction of progress in this world.

The Project of Supercoherence

At its plenary meeting of December 3 and 4, 2016, the Club of Budapest adopted the project of supercoherence as a core activity of its international network of twenty-three member clubs.* The project, codesigned by Club of Budapest Creative Program Director Anne-Marie Voorhoeve, is to function on the basis of a Charter of Values and Goals inspired by the following considerations.

1. The contemporary world is economically, socially, and ecologically unsustainable. It is nearing a critical bifurcation with two possible outcomes:
   - transformation into a sustainable economic, social, and ecological system,
   - breakdown into chaos and violence.

2. The required transformation concerns the basic *modus operandi* of the human community. Whether it occurs in time depends on the values and goals of those who critically influence the functioning of the global community; their values and goals need to shift from self-centered competition without regard to consequences to system-wide cooperation in view of the sustenance and flourishing of the whole system.

3. The indicated values and goals cannot and will not be dictated by a concert of big government and big business; they need to and can arise from the grass roots. Their effective spread and penetration can be ensured by groups of dedicated individuals who create spheres of sustainability in their own region and link up to create progressively larger spheres. Through this process even small groups can produce the "critical fluctuation" that, in a period of crisis and instability, decides the outcome of a bifurcation.

4. Recognizing that the flourishing of human life and by implication of all life on the planet is the supreme goal, the Club of Budapest adopts the project of creating Centers of Coherence through its network of national and regional clubs and linking

---

*See www.theclubofbudapest.com (accessed April 25, 2017).

the centers in the common purpose of contributing to the achievement of the supreme goal.

Implementation

The project of the Club of Budapest calls for the members of its international network to constitute themselves as Centers of SuperCoherence (CSCs) and adopt for their goal the creation of spheres of coherence in their country or region. The CSCs are to be linked among themselves by the international network of the club in a system of communication.

The project is not limited to the members of the network; these only serve as initial focal points and catalyzers. The purpose is to increase the level of coherence in the contemporary world, and this purpose is best served by creating CSCs in various countries and regions of the world and linking them in a system of contact and communication.

The supercoherence project calls on humanistic organizations the world over to begin functioning as CSCs. In practice this means that the centers undertake to:

- survey their own organization for levels of coherence: levels of contact, communication, and cooperation starting each on a personal level and then moving to interaction between themselves and among their own members and collaborators;
- survey the coherence of their organization with their social environment: levels of contact, communication, and cooperation with like-minded groups and organizations in their nation or region;
- survey the wider social coherence of their organization: contact, communication, and cooperation with groups and organizations working in other (social, political, economic, ecological) fields in their country or region;
- survey the ecological coherence of their organization: its respect for the integrity of the environment locally as well as globally;

- and assess the level of social and ecological coherence in their country or region and report the findings for the intention of decision makers as well as to the public at large.

The CSCs are to identify practical steps for increasing the level of social and ecological coherence in their country or region, as well as the coherence of their country or region with other countries and regions. This could be an important step in creating a critical mass of informed, committed, and effectively interlinked organizations—a network that could shift the aspiration for supercoherence from the level of concept and theory to that of commitment and practice.

The Need for Planetary Consciousness

The answer offered in this book suggests that the purpose of our existence in the world is not only our evolution toward supercoherence, but the evolution of our consciousness. The latter is an urgent and important issue: the consciousness of the modern age is far too narrow to respond to the challenges that face us. (In this context *consciousness* refers to the sphere of our intellectual/rational awareness as well as the range of our subconscious awareness.)

It is often said that we confront the problems of global society with the consciousness of a tribe, but that is not quite the case: the case is even worse. Tribal consciousness is deeper than the typical consciousness of the modern age. In the consciousness of modern people the dimension of instinct, insight, and spontaneous apprehension is missing; it has been dismissed as mere imagination, a carryover from mysticism and esoteric speculation. But the simile is apt inasmuch as tribal consciousness focuses on matters of the tribe and its immediate environment and leaves out of account the world beyond the range of its immediate concerns.

We need a wider and deeper, more planetary consciousness. We are an integral part of humanity, a system embedded in the biosphere, which is itself an integral element of the universe, but we treat the world around us as if it was an autonomous domain only externally and insignificantly related to the rest of the world. We treat the world as a mechanical array of independent elements, a system that is decomposable to its parts. We care only for the part that we perceive as ours, and do so at the expense of all other parts. This generates a wide range of conficts: economic, social, as well as cultural.

In regard to this issue, the first project of the Club of Budapest, launched over twenty years ago, acquires fresh relevance. The founding document of the club speaks of the need to evolve our consciousness, and the Club of Budapest has been dedicated to furthering this evolution. The following excerpt from "The Manifesto on the Spirit of Planetary Consciousness," the founding document of the club, drafted by Ervin Laszlo with the Dalai Lama, defines why we need to reach for a planetary consciousness and what we are to understand by this idea.

CALL FOR PLANETARY CONSCIOUSNESS

In most parts of the world, the real potential of human beings is sadly underdeveloped. The way children are raised depresses their faculties for learning and creativity; the way young people experience the struggle for material survival results in frustration and resentment. In adults this leads to a variety of compensatory, addictive, and compulsive behaviors. The result is the persistence of social and political oppression, economic warfare, cultural intolerance, crime, and disregard for the environment. Eliminating social and economic ills and frustrations calls for considerable socioeconomic development, and that is not possible without better education, information, and communication. These, however, are blocked by the absence of socioeconomic development, so that a vicious cycle is produced: underdevelopment creates frustration, and frustration, giving rise to defective behaviors, blocks development. This cycle must be broken at its point of greatest flexibility, and that is the development of the spirit and consciousness of human beings. Achieving this objective does not preempt the need for socioeconomic development with all its financial and technical resources, but calls for a parallel mission in the spiritual field. Unless people's spirit and consciousness evolves to the planetary dimension, the processes that stress the globalized society/nature system will intensify and create a shock wave that could jeopardize the entire transition toward a peaceful and cooperative global society. This would be a setback for humanity and a danger for everyone. Evolving human spirit and consciousness is the first vital cause shared by the whole of the human family.

In our world static stability is an illusion; the only permanence is in sustainable change and transformation. There is a constant need to guide the evolution of our societies so as to avoid breakdowns and progress toward a world where all people can live in peace, freedom, and dignity. Such guidance does not come from teachers and schools, not even from political and business leaders, though their commitment and role are important. Essentially and crucially, it comes from each person himself and herself. An individual endowed with planetary consciousness recognizes his or her role in the evolutionary process and acts responsibly in light of this perception. Each of us must start with himself or herself to evolve his or her consciousness to this planetary dimension; only then can we become responsible and effective agents of our society's change and transformation. *Planetary consciousness is the knowing as well as the feeling of the vital interdependence and essential oneness of humankind, and the conscious adoption of the ethics and the ethos that this entails.* (Italics added.)

The rationale for following the call to planetary consciousness is summed up in the pronouncements of some of the wisest individuals of our time.

A small and dedicated group of people, acting at the right time with the right idea, can change the world (Margaret Mead).

If it is to effectively change the world, the small group must act with a more evolved consciousness, because the problems of the world cannot be solved at the level of consciousness at which they were born (Albert Einstein).

A more evolved consciousness is not just to be advocated by the world-changing group; it must be practiced and lived. The group must become the consciousness it wishes to see in the world (Mahatma Gandhi).

A small and initially seemingly insignificant change in the consciousness of a small group can grow and spread rapidly in the world. Because consciousness is not confined to individual brains: it is nonlocal and one (Erwin Schrödinger).

Spreading and winning the adherence of people, the evolution of consciousness is an idea that is more powerful than any economic, military, or technological measure. There is nothing as powerful in the world as an idea whose time has come (Victor Hugo).

The call for planetary consciousness, issued twenty years ago, is as timely today as ever. The evolution of planetary consciousness is clearly the next stage in the evolution of humankind. Only this consciousness can reflect and respond to the reality of an emerging global society; it is the precondition of living in peace and well-being on the planet. We need a new consciousness to see the world and see ourselves in the world. Recognizing who we are and why we are in the world and evolving our consciousness merits our urgent attention.

7

The New Answers and the Purpose of Business

Dawna Jones

According to the new paradigm in Laszlo's books, the new bottom-line answer to the question *Why are we here?* is clear and even evident. We are here to transmit more and higher consciousness into the universe. The question raised in this contribution to the clarification of the meaning of the new answers concerns the world of business. What is the real purpose of business? Could it be that it is to contribute to the transmission of more and higher consciousness into the world?

Google the purpose of business and you'll see economist Milton Friedman's 1970s definition, "to maximize profit for shareholders." Reference.com states that the primary purpose is "to make money." Business visionary Peter Drucker's version, "to create and keep a customer," adds a wider focus with more beneficiaries. All three hold limited aspiration and restrict human potential from achieving what it is truly capable of achieving. Each purpose focuses on a part of the larger whole, while being employed as a decision-making principle, consciously or unconsciously. Because purpose provides compelling reason to exist, anchoring decision making at every level to the deeper purpose could inspire the evolution of business.

Maximizing profit for shareholders has been coined the "dumbest idea in the world" by Jack Welch, former CEO of General Electric, echoed by Forbes contributor Steve Denning, and for good reason. When executive compensation is tied to maximizing shareholder returns unethical behavior is the result. Wells Fargo is the most recent company to demonstrate how a focus on maximizing shareholder return created metrics resulting in consumer fraud.

Even without that effect, the beneficiaries of the company's purpose, defined narrowly, are restricted to a small part of the whole system. The health of employees, suppliers, and customers is willingly sacrificed, weakening the coherence and health of society and the environment. By focusing on a limiting purpose without wider benefit, incoherence on the inside of the company spreads to negatively impact the health of humanity and the ecological systems needed to sustain life. Through a limited consciousness, holding a limiting aspiration up as a purpose creates disengagement, causes stress-related illness, and breaches trust between business and society.

The trouble is that, viewed through limited consciousness, these dynamics are invisible, and so they continue to operate despite waving the signals. For workers in traditional companies operating under the "maximizing shareholder profit" definition, there is little sense of accomplishment. A 2009 study revealed that 60 percent of executives play video games for thirty minutes a day just to feel alive.

As game designer and futurist Jane McGonigal put it, reality is broken. Thankfully, this is the perfect time to create a different version.

How Can Business Live Up to Its Real Purpose?

Purpose guides the focus for decision making both consciously and unconsciously. It is also the inspirational beacon guiding expression of creativity and holds the key for engaging true talent. Transcending adversity, as expressed by the massive problems humanity is now facing, is where the beacon is located. Getting from where we are today to attaining the goal that business can be of benefit to the world on multiple levels—ecologically, societally, and economically, for instance—requires the evolution of consciousness in the world of business.

The choice to step forward into a higher level of leadership is nearing the place of no choice. All sizes of companies are failing fast, with longevity down to ten years and continuing to drop. Upstart startups and innovators are developing products and services that are disrupting and replacing business models, leaving large traditionally managed companies under pressure. The suffocating notion of business's purpose as "to make money" is being replaced by more ambitious reasons for business's existence.

Ervin Laszlo describes the process of bifurcation where the linear trajectory of evolution reaches a pivotal choice: transform or collapse. He writes there are laws of complexity, including laws of the evolution of complexity, that apply to all systems. A nonlinear system-level transformation called a bifurcation is one such law. It occurs in the evolutionary trajectory of nearly all complex systems, and it spells the end of some systems and the transformation of others. In business terms, the collapse of the linear system places business leadership in the crux of choice: transform or collapse. Companies grasping for familiar linear strategies to control the effect of disruptive exponential technologies choose the latter, unwittingly.

Writer, inventor, and designer Buckminster Fuller said, "You never change things by fighting the existing reality. To change something, build a new model that makes the existing model obsolete." The new models are replacing the existing models at warp speed. In the energy sector, Tony Seba, Silicon Valley entrepreneur and lecturer at Stanford University, forecasts that unsubsidized solar power will replace all fossil fuels by 2030. IKEA is already using 100 percent solar. Electric motors, for example, are five times more efficient than the internal combustion engine, converting 85 to 95 percent of energy into kinetic motion. Operating costs are significantly lower. Battery range is increasing and prices are dropping to bring the electric car to the same price range as the gasoline-powered vehicle by 2023 or earlier. These and other changes render linear business thinking useless as the systemic effects ripple throughout all relationships in the interconnected system. With a focus on the parts, rather than seeing the integrated whole, companies resort to running faster in the hope that complexity is a cycle on the linear runway of conventional business thinking. Companies can use a higher purpose to rise above well-worn patterns. However, this will take a higher level of leadership's consciousness.

On the evolutionary path are companies that focus on Drucker's definition: to create and keep customers. It is goal of wider benefit, but is it big enough to lead to further evolution?

As costs drop hastened by reducing waste, efficiency increases. New technologies like robotics and smart machines are already replacing humans and are triggering fear. Can a creative response consistent with an evolutionary mind-set replace fear? Can trust in one's talent replace seeking security found in job certainty? Transforming dormant human potential into revitalized coherence activates the evolutionary leap to redesign or reinvent before crisis leaves no other option. Thinking big is the catalyst.

Evolving human consciousness is about tapping into the creative adaptive spirit biologically wired into each human being. "We actually contain a built-in ability to rise above restriction, incapacity, or limitation and, as a result of this ability, possess a vital adaptive spirit that we have not yet fully accessed. While this ability can lead us to transcendence, paradoxically it can lead also to violence, our longing for transcendence arises from our intuitive sensing of this adaptive potential and our violence arises from our failure to develop it."[*] Adversity and massive problems turn into opportunities for business.

The bigger the challenge, the better. Each seemingly insurmountable

---

[*]Joseph Chilton Pearce, *Biology of Transcendence: A Blueprint of the Human Spirit* (Rochester, Vt.: Park Street Press, 2004), 2.

problem is an invitation to expand consciousness and evolve: personally, organizationally, societally, and globally. Co-creating a new version of reality is the challenge we collectively face as a species.

Moonshot 10X goals, goals that reach beyond the capability of any one person or company, are the kinds of goals that activate intuitively creative responses. Google's mind-set and Peter Diamandis's XPrize Foundation demonstrate the value of thinking big. In contrast, "Achieve quarterly targets" is uninspiring.

Can business go beyond profit as its purpose? It must or it will abjectly fail. Is the current purpose of business in alignment with transmitting higher consciousness into the world? And if it were, what would business then look like?

Novo Nordisk is one of several large, global, publicly traded companies over a hundred years old that have embedded life's principles into their cultural DNA.* Their deeper purpose (or goal) is to achieve systemic health. Systemic health is a unifying value and so it serves as an organizing principle for decision making at all levels in the company. Because achieving systemic health benefits all life, beyond humanity, the entire workforce is inspired to contribute love for their work, which is more than their intellectual and creative talent.

Applying the New Map of Consciousness Going Forward

Consciousness is a word that is too frequently tucked away in the "new age" file, thereby bypassing the chance to make sense of the critical point now facing humanity. Ervin Laszlo's practical definition describes consciousness as "the determining factor of how we see the world, of who we are, what the world is, and what we can do in the world. It is the mindset—the totality of preassumptions, assumptions, intuitions,

---

*Joseph H. Bragdon, *Companies That Mimic Life: Leaders of the Emerging Corporate Renaissance* (Saltaire, UK: Greenleaf Publishing, 2016).

and information about the world, each other, the possibilities, dangers, and opportunities."* To bridge linear thinking and limited consciousness with an up-to-date worldview means accessing more than one way of thinking and perceiving. *"If we try to pick out any thing by itself, we find it hitched to every thing else in the universe"* (John Muir, 1911).

The new map of consciousness provides insights and a photosensitive lens to guide the evolution of consciousness in business. Three areas converge to show the way.

1. *Understanding how complex adaptive systems work.* Traditional business mind-sets work with linear and intellectual assumptions about how things get done. It is a narrow view. Alternatively, understanding the characteristics of complex adaptive systems speeds transformation while restoring a sense of playfulness. The joy of collectively accomplishing a challenging task fuels accomplishment. Interconnectivity and coherence are inseparable in living systems. In companies, this is likewise true and is expressed through networks of interaction.

2. *Engaging the power of the human spirit.* In business terms, the human spirit is defined as the ability to "take your whole self to work." Doing so means that organizations are designed from the outset to receive creative contribution openly, being self-managed freedom-centric companies. Alternatively, organizations can be redesigned, a more difficult challenge made easier by incorporating the principles of complex systems. Power is redistributed without dismantling hierarchy through participatory process and responsible leadership. Parallel to organizational redesign is the human side: it is to evolve to a full-spectrum consciousness comfortable with co-creating in a peer-to-peer network without leaning on authority.

---

*Podcast conversation: http://insighttoaction.libsyn.com/-toward-a-new-business-leadership-consciousness (accessed April 27, 2017).

On a wider scale, the role of business in society is being redefined by companies bold enough to benefit the world. Initiative, autonomy, responsible leadership, habitual learning, and a sense of control—even when faced with uncertainty—are sourced in the human spirit. Psychologically, humans prefer to wait for a crisis. But when operating intuitively, vision and foresight activate action before the advent of crisis.

3. *Growing self-realization, organizational company-wide consciousness.* The skills associated with working in a highly networked autonomous organization are far more sophisticated than those required for a command-and-control style organization. At a personal and collective level, aiming for a massive, meaningful purpose restores connection and belonging: two attributes that high-performance companies understand. The ripple effect means fewer people accomplish higher quality results with greater awareness of consequence and of the impact of their contribution. Better business is bound to result.

From the limited consciousness residing in approximately 85 percent of U.S. leaders, it is not possible to see the natural networks of performance or the impact of decisions on customers, employees, and societal and ecological health. Nor is it possible to observe the interaction of the formal structure with the informal social and emotional networks that connect and form a coherent field. Fear is undetectable but fully operational. For business to evolve to a higher purpose, the entire worldview must widen and deepen to reveal what makes the system coherent and stable even while transforming. Trust and confidence must replace fear.

The three lenses below offer ways for business decision makers to appreciate what lies beyond the boundary of existing knowledge and prevailing assumptions about how the world works. Pivoting worldview enables actions that support coherent relationships and ripple throughout the entire dynamic: inside, as well as beyond the boundaries of the company. Not to be confused with systems thinking—a mental track—the three lenses offered here provide a means to sense and see the vibration of energy-driving performance, reputation, compassion, and care. In physics, the definition of energy is "ability to perform work." The energy that is expressed or repressed signals the given consciousness of the company and may become the dynamic platform for its growth.

Lens 1: Understanding How Adaptive Systems Work— The Lens of Networks

Viewed from the dominant business consciousness, most of what drives performance in companies remains unseen. To achieve coherence inside a company between the informal interactions necessitates observing the emotional and social networks that power performance. Networks exist in all organizations, in hierarchies with centralized decision-making authority, as well as in self-managed or matrix structures where leadership and power are distributed.

Performance runs on networks no matter whether it is in a matrix organization, networked by design, or in a traditional hierarchy. High-performance networks are coherently organized around a shared aspiration. Ego-driven networks operate in companies at a limited level of consciousness where there is a greater need for psychological safety and where the focus is on self-preservation or gain. The first is driven by "We"; the second by "Me."

Companies that manage by applying the principles of a living system are aware of the importance of the networks to shorten response time between issue identification and response. On the other hand, companies and management that still rely on authority for control cannot perceive interpersonal network dynamics. The narrow mental focus blinds depth perception pivotal for discerning what is powering and sustaining performance.

In 2005, Hewlett Packard mapped who was involved in accomplishing a goal that mattered to the team. They were seeking to understand what sustained phenomenal performance in the ink-jet division. The answer was Joy.

The research found

> that the people involved in the accomplishment were rarely identified by the organizational chart. While the formal organization chart was designed to control, performance was driven by networks of people and employees (including managers that transcended the formal organization chart) . . . when we looked at some of the networks we studied in Hewlett Packard and the U.S. government, we discovered that there were people from almost every activity in the company at almost every level working on a particular objective . . . including customers, competitors, suppliers, and just about anybody you could think of, who needed to be a part of that accomplishment.*

Understanding networks gives managers a more meaningful role to play and a more powerful contribution, but it takes a desire to become more self-aware at the emotional level in order to let go of the intent to control others.

After Nick Zeniuk, a former Ford executive and mentor, described networks to a manager in a large European company, the manager responded with: "Wow, so this is the way we get things done. I didn't realize that this was the way we were doing things. Now we can improve it, redesign it, and make it better." Her response marks the microsecond where mindfulness, the ability to self-regulate and manage your response, is the door to knowing whether to act on habit or wait and watch.

Zeniuk explained, "The moment you decide to redesign the informal networks is the moment that the network goes underground and disappears. These networks are self-organizing. You cannot design

---

*Conversation with Nick Zeniuk, former Ford executive, in podcast episode "Follow the Joy," published on January 21, 2008, www.management-issues.com/podcasts/48/follow-the-joy and https://itunes.apple.com/ca/podcast/evolutionary-provocateur/id1006636782?mt=2 EP82.

them; they emerge around shared inspiring goals." Functioning effectively in complex systems requires being able to maintain a high level of conscious oversight. Otherwise, managers and executives exercising authority are tempted to interfere by formalizing what works better as an informal way of getting things done.

Networks follow a sociobiological pathway and so are not detectable through surveys and conventional intellectually familiar tools for isolating performance qualities. To observe networks, to see patterns, to detect the deep dynamics, sense of joy, belonging, and connection requires depth perception. Being a compassionate observer of the currents of energy, of focus, of care, of whatever matters, illuminates the hidden leverage points where a small effort creates a system-wide effect. As the networks inside a company become visible, the capacity to see into societal and ecological interrelationships is a natural extension. With a wider view of the landscape, there are fewer surprises. When surprises do happen, everyone in the company is better prepared.

Lens 2: The Lens of Emotion and the Human Spirit

Lens 2 shines the light on why organizations only get incremental change when the intention was for something more ambitious. To evolve leadership consciousness, concurrent with economic restructuring and ecological debt repayment, a more thoughtful and responsible response is required. Anyone who has ever orchestrated organizational change initiatives will have sensed what is most often attributed to "resistance to change" and human fear—a psychological view.

Obviously, there is more to it than that. After observing what felt like an energetic holding pattern in team and organizational dynamics, I was looking for insight and preferably some science to make sense of my intuition. In 2009, Rollin McCraty, Ph.D., research scientist from the HeartMath Institute, and I were talking about energetic sensitivity and intuition. It was a logical extension to ask about organizational energetic patterns (blueprints). McCraty referenced the work of Raymond Bradley, Ph.D., who studied the emotional relationships in several organizations.

Two important factors emerged:

1. whether people felt they had power; or
2. whether others had power over them.

McCraty explained, "Then he [Bradley] created energetic maps based on these two main factors: affect, positive or negative emotional relationship, and who it was agreed had more power or structure. These maps were very telling about the organization. He used the word *coherent* to describe very beautiful coherent relationships. He could predict the long-term survivability or success, even academic success, based on this network of relationship." Eventually Bradley produced a literal quantum holographic model to explain the data. But what is the practical value and relevance of this?

Paying attention to the quality of trust in workplace relationships, how conflict is utilized, how diversity is employed, and the emotional tone in communication, are critical success factors. When negative or toxic relations are considered acceptable, an incoherent field is generated. Once it is set in place at the energetic level, it is difficult to change.

These fields are intuitively observable in company dynamics and are revealed in the way failure is handled. A high-level trust environment uses failure to build something better the next time. A low-trust environment creates depression and stress-related illness. McCraty explained that what I was detecting is a real energetic field. "We can't measure the field directly. But we can't say these organizational fields are not real because we can't measure them. Even a well-established concept like an electrical field has never been measured. No one has ever measured a field of any kind directly, but we can measure the effects of the field."

When a company tries to reinvent itself and the energetic fields are not taken into account, incremental results are the only possible outcome. The company brain, like the human brain, will revert back to what is known and comfortable.

In business, the effects of the field show up in results: in the health of employees, degree of workplace trust, financial health, in valued relationships with customers, and in ethical decision making. The critical importance of high-quality conversations, capable of sustaining trust by tackling the difficult conversations, is both a vehicle and a metric for progress. *If you want to navigate complex organizational interactions, follow the emotional energy—its frequency and its vibration.*

The energetic fields McCraty describes express the consciousness embodied in the organizational mind-set. Emotional energy vibrates at different frequencies. Anger has a perceptively different vibration than joy, for instance. Whether or not you are aware of how you are feeling, the workspace indiscriminately holds the mix of energies. The overall emotional vibe has a direct influence on decision making, risk taking, health, and the kind of customers and shareholders a company attracts. For instance, companies holding a profit-driven mind-set attract short-term investors in for the quick turnaround between investment and return. A company evolving toward a higher consciousness, guided by a higher purpose and worldview, balances long with short term. It also attracts shareholders who see the value of being patient for considerably higher returns.

To illustrate, the CEO of Unilever, Paul Polman, has made it clear to shareholders and prospective investors where the company stands. He sees Unilever as an agent for tackling child poverty, climate change, and other societal issues. When asked by an investor seeking short-term results, he took a clear stand: "Unilever has been around for 100-plus years. We want to be around for several hundred more years. So if you buy into this long-term value creation model, which is equitable, which is shared, which is sustainable, then come and invest with us. If you don't buy into this, I respect you as a human being but don't put your money in our company."* The integrity and principled leadership that goes with a company confident in its focus on higher purpose and

---

*Conversation with Joseph Bragdon, author of *Companies that Mimic Life,* on the Insight to Action podcast: http://insighttoaction.libsyn.com/how-do-companies-that-mimic -life-perform-financially (accessed April 27, 2017).

talent diffuses throughout the product line and is a source of pride for employees. Unilever is a global company steadily evolving its state of consciousness.

The point for business is that meaning supersedes metrics. Rather than capital assets, high-quality, trustworthy relationships between every point on the constellation, from employee to customer to society, create value. Quantity is replaced by the quality of the relationship and the capacity to work with diversity in a fluid, natural way. Restoring coherence to the role of business in society requires that business accept responsibility for the impact of its decisions on the natural world, on society, and on the well-being of employees and customers. Companies operating with this worldview far outperform their peers financially, not because they aim for profit, but because they aim for a higher purpose.

Epigenetics and Workplace Health

Epigenetics literally means "over or above genes." For evolutionary purposes, awareness of epigenetics provides business leaders with a critical insight into creating healthy decision making and ethical environments. The purpose of including epigenetics and emotional awareness in a discussion around consciousness is to shake up the assumption that business decision making is rational. If it were, businesses would have evolved their place in society much sooner. Business decisions are made based on beliefs. Rigidly held beliefs block agility, impede innovation, and lead a company to extinction by default. A brief discussion of emotion and epigenetics sheds light on the fallacy of rationality as the right means for decision making with the intent to open the door to a more holistic worldview.

The sea of data in and around us is saturated with emotions emanating from our thinking, from the subconscious, from interpersonal communication, or held as energetic blueprints in the workplace environment. Without conscious awareness, the data tells our brain whether we are safe or at risk. When fear, anger, or other discordant emotions typify the environment, biologically we go into protection mode and the need for safety dominates unconscious decision making. Risk taking shrinks and decisions are motivated by self-preservation. Ethics are compromised. But where trust and a sense of support and belonging provide stability, creativity flourishes. Collectively, we can co-create new solutions while navigating chaos.

The emotional mood in the environment is an expression of how we are connected to one another. Progressively becoming more aware of the coherence or de-coherence in the emotional and social environment allows intentional choice regarding the quality of life. Feeling a sense of control over our life (personal power) is a determinant of health on its own, but it is also essential for leading life in a healthy and positive direction, particularly in climates of uncertainty.

Toxic workplaces or relationships compromise physical and organizational health. Mental stamina decreases; creativity is inaccessible given the high-alert protective stance needed to get through the day. Ultimately a de-coherent energetic field is established and can become entrenched. De-coherent energy is often palpable to energy-sensitive persons entering the building. Even customers are aware that something is not right. At a personal level, simply being more aware of the impact of personal behavior can serve to improve relationships.

In toxic work environments the body goes into protection mode. Apart from being unhealthy, fear transmitted through communication or management style shrinks decision options down to one, even though more are available. Fear-based decisions motivated by protecting reputation or social status lead to unethical decisions. At the limited end of consciousness, these decisions are unconsciously motivated by the desire for personal gain or to restore fairness in what is perceived to be an unfair relationship. Management styles and the application of the wrong metrics in many large command-and-control environments create unethical behavior. In contrast, trust-based environments, characterized by non-judgmental awareness, provide access to diverse thinking and creative solutions. Adding understanding through neuroscience, biology, and physics provides individuals and teams valid tools for improving the quotient of trust and of quality communication. At the leadership level, contextual awareness and the understanding of epigenetics is pivotal to gaining insight, developing multiple options, and holding a system-wide overview.

The energetic fields, consciously imperceptible but perceived by employees and customers at minimum, are set through emotional patterns. Fear and stress in the workplace create stress-related illness costing business in the United States an estimated $190 billion per year—and a needless cost in quality of life. Though companies might stigmatize stress-related illness, the blame avoids paying attention to the health of the workplace as a distributed and shared responsibility.

Over three years following the 9/11 attacks, University of California–Irvine research found acute stress responses to the attacks on the World Trade Center. Individuals worried about terrorism were three to four times more likely to report a doctor-diagnosed heart problem two to three years after the attacks. If fear is entrenched in the workplace, or in the social or communicational environment, stress-related illness results. Similarly, if management tells or directs talent how to contribute, removing a sense of control and autonomy, biologically, depression or aggression results. The upshot is that business is costing itself money and a high cost in quality of life by resisting evolution.

When it comes to evolving consciousness, fear presents the contrast to joy and a choice to restore trust in human ingenuity. Great workplace cultures are created when there is diligent attention given to creating positive connection even when facing conflict, rather than watching it spiral down to a win-lose dynamic.

At a personal level the sense of control to regulate emotions, to be aware of the impact of feelings and interactions on personal and working relationships, is part of our self-leadership. The more elevated our consciousness, the more we can detect and adjust the spirit and emotional tenor in the workplace environment and adopt inventive ways to reset emotional patterns. Attending to health in relationships has noticeable effects on risk taking, psychological safety, and behavior; on what we choose and what we believe is possible.

Lens 3: Evolving through Transcendence—
The Lens of Tapping In to Deep Adaptive Spirit

Failure to rise above systemic ruts is rationalized by the belief that it takes time, and since business professionals are too busy for that, routine patterns run daily life. When companies do try to change, there is a tendency to select from linear processes ineffective in complex nonlinear systems. Consequently more effort is expended and results are incremental at best. In nonlinear systems, small emotional and spiritual disturbances, novel and creative, can quickly ripple throughout the larger system. An entire city has been transformed by applying the principles of complex adaptive systems. Two visionary mayors, Antanas Mockus and Enrique Peñalosa, transformed Bogota, Columbia, from near anarchy to a revitalized city over a ten-year period. Employing mime artists and other creative methods, even traffic patterns have changed. Documentary filmmaker Andreas Dalsgaard explains the success of Mockus and Peñalosa as leaders:

> The real secret behind Mockus and Peñalosa's success is that they are two people characterized by extreme honesty and integrity in everything they do. They are two leaders who have the necessary courage to stay true to their visions, even when the opinion polls go against them. Unlike other politicians who are controlled by strategies and tactics, they have not been driven by a lust for power, only by their ideas and philosophies. And if there is a lesson to be learned by their story, it must be that the change they have managed to bring about could never have come from the traditional political system. It could only have come from the outside.

Bogota's population in 2005 was between six and seven million. If a company of less than five hundred thousand employees is bold enough to tackle a large global issue that inspires people and matters, then transforms itself through the vehicle of intelligent play, wholescale change will take far less time. Employing methods that engage creative curiosity is a clear opening for transformation and evolution.

Polarity is another opening for transcendence, for rising above duality and gaining collective strength. On the other hand, without emotional awareness polarity results in feeling emotionally victimized when outcomes do not match expectations. Brexit, where Brits voted to leave the European Union, and the election of Donald Trump have triggered more fear than hope and catalyzed political activism by people eager to design more inclusive alternatives.

Rising above polarity to build compassion and empathy includes actions such as not judging people who shaped the outcome as right or wrong, as smart or stupid, but remaining curious about the *why* of the decision, detaching from drama and blaming others. At the meta-skill level the capacity to self-observe, self-reflect, and self-correct in the present moment sums up what conscious leaders do better than those who are unknowingly driven by systemic beliefs. And if the polarizing event instigates the bigger questions, then this book, dedicated to exploring who we are and why we are here, has more than served its purpose.

We are in the period of the sixth mass extinction—and the first to be caused by human activity. "Our global society has started to destroy species of other organisms at an accelerating rate, initiating a mass extinction episode unparalleled for 65 million years."* We are connected from the far reaches of the cosmos to the center of the planet. It is no accident that business is being called to evolve, not just to keep pace but to lead the way. Institutional reform—political, health, and educational—will take longer.

The Need to Evolve

What could happen should business leaders make the leap to the new paradigm of purpose in business? How might business reconnect to

---

*Gerardo Ceballos et al, "Accelerated Modern Human–Induced Species Losses: Entering the Sixth Mass Extinction," *Science Advances* 1, no. 5 (June 19, 2015).

being a responsible global citizen? How might business regain coherence with the planet and the people it serves?

New organizational designs are emerging from startups and companies reinventing themselves. The new designs share similar organizing principles such as transparency, autonomy, and leaders in every person. For those in the company, self-realization is the path to achieve personal coherence and, ultimately, organizational coherence. To be fluently and fluidly responsive to emergent external conditions is the aim.

Can companies reach enough flexibility to mimic the fluid coordinated movement of a flock of birds, responding as a single intelligent cloud? Change is detected through a signal, instantaneously communicated; the entire flock moves as one.

Italian business innovation company Cocoon Projects has designed itself around fluid dynamics, calling their governance LiquidO (O = organization). Embedded in the company's design is *stigmergy*—indirect coordination arising from modifying and sensing the local environment where one signal builds on another—the same communication pathway employed by anthills, beehives, and in chess. Stigmergy is an enabler for self-organization in complex multicellular organisms employing both signals and cues to communicate. Other nature-inspired organizational designs provide an organizing structure; higher purpose inspires the work.

After the 2008 banking crisis it became clear that the banks were guilty of fraudulent activity. Not only did the banks not hold themselves accountable for their actions, government did not either. Author Eugene Linden asked: "If you are overextended who do you turn to? Your broker or a vampire bat?" Until the financial sector progresses, choose the bat. Vampire bats store extra blood (a liquid asset) for a sick bat to call on when needed. Inspiration comes in many forms.

What if banks contributed to the health of society? By restoring economic health to neighborhoods, for instance? Rising above multiple layers of systemic resistance, there are better solutions with benefit to all.

The question is how we can live up to the purpose of our existence.

My answer is that it is up to each person, each reader of this book. For the world of business, the answer is sourced in the aspiration of the company: either to be a business of human benefit or to add to humanity's ecological and societal overdraft.

Though the challenges seem insuperable, complex challenges call forth the deepest in human potential. Achieving seemingly unsolvable, impossible challenges builds strength, character, and empathy by accessing the creative, adaptive spirit that is wired into each person. Without that challenge the innate, adaptive spirit would lie dormant. Aggression and violence reveal failure. Yet, those dark moments exist to provide contrast to experience the lightheartedness of pure joy—doing work that regenerates and restores vitality in humanity and the planet. The invitation is to evolve toward embracing a unifying view of life both at work and at home, living with multicultural perspectives, compassion, and oneness. Socrates said that the secret of change is to focus all our energy, not on fighting the old, but on building the new. Bringing awareness to where energy is focused and developing subtle sensitivity serves as the compass for finding the way, as indigenous navigators knew when traversing the open ocean. Navigating the new map of consciousness to actively participate in the unifying purpose of life calls for achieving individual coherence; coherence within a company, societally, and globally, while bringing hope to the possibility that biosphere consciousness is attainable.

Technology is a tool and not a replacement for the choice to manifest the innate intelligence that connects us to each other and is the ultimate goal of evolution. Designing alternatives to worn-out institutions and developing entrepreneurial initiatives can replace fear of job loss and renew a sense of control. The challenge before us is not for the great minds of the world to resolve, but for the great hearts in each person to do so.

You may ask: "Why me? Why should I accept responsibility for activating the deeper purpose of business?" Consider that a beam of sunlight splits into red, orange, yellow, green, blue, indigo, violet. If each person represented one of those colors, all the colors could form light. Each person would bring a unique spark to the conversation, to the ultimate goal of bringing humanity into alignment with the intelligence of the cosmos. You are a unique human being, with all its innate potentials. *Why not you?*

8

The New Answers and the Wisdom Traditions

Shamik Desai

Who are we? Our essential nature is identical to the essential nature of the cosmos—pure consciousness, or love, or spirit. *Why are we here?* According to all the major wisdom traditions, we are here to access, embody, and transmit this divine consciousness into the world until material reality is made sacred—that is, until cosmic consciousness reigns on Earth or, alternatively, until ultimate reality (God) returns to its original form (of infinite oneness).

The question arises: Why did God manifest the universe at all? The great religions hint regarding the meaning of existence: the One Source yearns deeply to know and love itself. Hence, it has manifested in a material form (at a higher band of vibration, in Laszlo's language) with the intention of realizing itself through a process of respiritualization—to which we ourselves contribute.* The economists—the secular priests of today—insist that we humans are utility-maximizing creatures. If this dubious claim is true, then humanity's deepest desire and highest

---

*Shamik Desai, Afterword in Ervin Laszlo's *What Is Reality? The New Map of Cosmos and Consciousness* (New York: SelectBooks, 2016), 265–66.

purpose is to maximize spiritual fruits—to maximize the consciousness transmitted into the universe. It is in the currency of consciousness that we find our deepest security, wisdom, and joy, that we fulfill our ultimate destinies and enrich God Itself.

The symbols and semantics of the various wisdom traditions may vary, but they speak to us with one voice. When we peek behind the symbolic garb and uncover the gist of the religious messages, we find that their central teachings are strikingly unified—teachings that are also strikingly consonant with the findings emerging at the leading edge of modern science. Common themes emerge across the traditions. The world is more integrated than it appears, and also, being suffused with loving intelligence, more benign. Our task is to become aware of this wholeness and warmth and to amplify and accelerate it.

We divinize the world by being "good." But what does this mean exactly? Laszlo would define goodness as that which promotes complexity, coherence, and compassion. The religions define goodness as that which is aligned with God's will—meaning that which is aligned with the true nature of the intelligence of the cosmos. We are good when we orient our lives to, and harmonize our world with, the oneness of reality—simply put, when we act selflessly.

Ethical precepts are decreed universally across traditions both to preserve moral order at a social and cosmic level and to instill virtue in the hearts of individuals. These might be thought of as the two prerequisites to freedom, as the former prevents us from being enslaved by "outer demons" and the latter by "inner demons." Hence the original Latin meaning of *virtue* as "power"—for the man without virtue is never free. Once we have mastered ethics (the Ten Commandments of Judaism, the Eightfold Path of Buddhism, the Five Pillars of Islam, and so on), we have mastered ourselves. We now treat others and ourselves justly, and we are ready to move to a higher (or perhaps deeper) level of our spiritual journey: love.

As we open ourselves to God and receive God's all-embracing love, as we allow this spiritual essence to flow into our hearts and, from our hearts, into the hearts of others, we receive and transmit consciousness into space and time (as Laszlo puts it). One might argue there is a spiritual stage even beyond transmitting love: transmitting the right *kind* of love—the particular consciousness that only we, uniquely, can transmit—by leading a life of purpose. In other words, by allowing our loving acts to fall neatly into the groove of a larger and coherent theme that resonates with our individual souls and enhances humanity and the cosmos. When we have visions of oneness and are flowing with purpose, when our lives are a response to a felt higher calling, we are leading lives of perfect coherence with ultimate reality.

This final, utopian stage, in which we become a fully conscious world (a global "community of called ones"*)—when every being is doing precisely what she was meant to do, unfolding, freely and willfully, toward her destiny—the entire planet becomes an interconnected field of oneness, enveloping the world in the highest vibrations of consciousness.† ("We can be at peace," the Mahayana Buddhists tell us, "because a boundless power draws . . . everything to its appointed goal."‡) When every particle has actualized its unique potentiality, when the earth has risen to the level of cosmic consciousness, we find something very similar to "the kingdom of heaven on earth" of the Judeo-Christians or the Omega Point of Teilhard. As our impulse to self-realize is woven into the fabric of Nature, our collective self-realization is perhaps ultimate reality realizing itself. By responding to the teleological tug of Nature, we co-create the divine. Hence, God—the Infinite—is both the beginning and end goal of history, the primal cause and effect: "I am the Alpha and the Omega" (Revelation 22:13).

Thus, evolution itself is the ultimate test of goodness and humanity's highest collective purpose. According to Laszlo, we evolve to trans-

---

*Shamik Desai, *See 2020: The World through Troogol Glass,* 298 (unpublished manuscript).

†*Highest* is used here metaphorically; technically, it is the *lowest* frequency band of vibrations.

‡Huston Smith, *The World's Religions* (New York: HarperCollins, 1991), 123.

mit higher consciousness into space and time. But what is the specific manner in which we do so? The world's wisdom traditions offer us two broad sets of answers. The Semitic (or Western) faiths—Judaism, Christianity, and Islam—call on adherents to embody, to *materialize,* the one truth, making heaven and earth one. The world itself, once sacralized through acts of righteousness, is to become a shoreless ocean of brotherly love: a Philadelphia, a Zion, a Promised Land. In the Eastern traditions—Hinduism, Buddhism, Taoism—followers are asked to reorient their perception of reality by peeling back the layers of illusion (*maya*) until the nondual core of consciousness (*advaita*) behind the veil of appearances is laid bare.* (The variation in emphasis can be seen when we consider that the concept of evil, in Hindu doctrine, is most accurately translated as *avidya,* or ignorance.) Chuang-tzu, disciple of Lao-tzu, spoke in a similar vein: "Let go of all your assumptions, and the world will make perfect sense."† While divine union (through spiritual evolution) is the ultimate destination of both sets of traditions, the paths are different.

Unlike in the West, the notion of preemptively and progressively "fixing" the world is alien to Eastern impulses: the Tao Te Ching notes, *"Those who would take over the earth and shape it to their will, Never, I notice, succeed"* (chapter 29), while the Bible states, "Let man have dominion over . . . the earth" (Genesis 1:26)—though it *is* one's duty in the East to *preserve* moral order when it is threatened. The Eastern traditions enjoin us to change our attitude to the world; the Western ones to change the world itself—to bend the arc of history rather than transcending history. The Western traditions urge us to supplant Nature's role, in a sense, and effect physical evolution ourselves by building increasingly complex and coherent social systems that culminate at the end of time. In the process, we evolve mentally— toward empathy, love, and oneness. For the Easterner, man's eternal,

---

*Shamik Desai, Afterword in Ervin Laszlo's *What is Reality?* 266–67.
†Roger Housden, *Dropping the Struggle: Seven Ways to Love the Life You Have* (Novato, Calif.: New World Library, 2016), 40.

inevitable cycling round the *kalachakra,* or wheel of time, renders the notion of improving on Nature to be foolish and futile. He focuses inward, on his own mental (or conscious) evolution, and physical evolution transpires by default.

But whether we uplift the world by uplifting ourselves or uplift ourselves by uplifting the world, the end goal is the same. As Laszlo noted: "These parallel processes of evolution are not separate: they unfold as aspects of a whole . . . the purpose of evolution in this universe is one . . . [it is] toward supercoherence [and] oneness." As we shall see below in the survey of world religions, mental and physical evolution are indeed inextricably intertwined—as they drive toward Teilhard's Omega Point of cohering humanity into an organic One.

While the central conclusions reached by the various wisdom traditions are almost identical (and also square with Laszlo's scientific hypotheses: we are clusters of vibrations, points of consciousness in an all-pervading field of consciousness; our purpose is to realize this consciousness and transmit it to others; and in so doing, to enable the intelligence of the cosmos to realize itself), each specific religion has its own tone and texture, and urges its disciples to transmit consciousness in its own style and manner.

Hinduism

In Hinduism, arguably the fountainhead of Eastern traditions, we transmit consciousness into space and time with a reverence for *duty*—something every Hindu must perform. The concept of God (or Brahman) in Hinduism translates most accurately as undifferentiated consciousness. Brahman is at once transcendent and immanent. "It is the ultimate reality from which everything proceeds and to which everything returns—the essence of the universe and simultaneously that of which the universe is merely an expression . . . the enveloping, all-creating consciousness that enfolds all that is."* (This sounds very

---

*Felipe Fernandez-Armesto, *Truth: A History and a Guide for the Perplexed* (New York: St. Martin's Press, 1997), 39–40.

similar to Laszlo's *logos* and Akashic field.) Each of us is an emanation of this ultimate reality like a spark of fire, or a ripple on an ocean's surface: "Atman (the individual soul) is Brahman (the Universal Soul)" is a Hindu catchphrase. Hindus must access, elevate, and permeate this divine consciousness by performing their duty, or *dharma:* the selfless, dispassionate execution of one's highest calling (which, in Hinduism, is delineated by one's life station). In this tradition, we each dispatch our roles like actors in a grand cosmic drama (*lila*).

The yogi's calling is to withdraw from the world of the senses and, through meditation (one-pointedness of mind), fuse his consciousness with the divine consciousness, achieving union with God. For those less fortunate souls who must continue to engage with the world, they are to execute their obligations with equanimity and disinterestedness. With such an orientation, each of us becomes a yogi, shifting our identification from atman to Brahman and renouncing all fruits save spiritual fruits. As the Hindu holy book tells us, "To work you have the right, but not to the fruits thereof" (Bhagavad Gita 2:47).

Hindus might liken "duty," then, to a consciousness-elevating and -transmitting vehicle. As we dispatch our duty in service of our loved ones, we accumulate positive merit, or good karma, and ascend the ladder of consciousness over successive lives, until our ego dissolves, and the individual soul merges with the Universal Soul, releasing us from the cycle of life (*moksha*). (At the end of each life cycle of the universe itself—defined as a *kalpa* of 8.64 billion years—everyone has ascended the ladder and all beings are perfectly enlightened. The apex of evolution has been achieved: Brahman has realized Itself and returned to its unmanifest state.) So in this process of humanity enacting its collective duty, Brahman evolves, experiencing joy and delight as it watches the cosmic drama unfold.

Buddhism

Buddhism teaches us to be *mindful* and *compassionate*—states through which we transmit consciousness into the world of space and time. All life is filled with *dukkha,* or suffering; we must endure it by transcending it, Buddha was convinced. As in Hinduism, freedom from ignorance and suffering is to be found through a shift in perception, in seeing things as they truly are—as Buddha put it, seeing things in their "suchness" ("*tathata*—the unique and ultimate reality which transcends all distinctions"*). We do this by entering into the deepest states of concentration and meditation—of mindfulness.

For Buddhists, like Hindus, the key to worldly living is purity of intention. Of the Eightfold Path—right views, right intent, right speech, right conduct, right livelihood, right effort, right concentration, right mindfulness—almost none relate to outcomes. We thus remain immune to fear and favor and extinguish *tanha*—the desire for selfish fulfillment. Hence, by not attaching ourselves to outcomes, or even classifying them as good or bad, we transcend the material world through our mindfulness.

In mindful modes, we "wake up" (enter into a state of nirvana, or ego-extinguishment) and realize that no material form is real or permanent; that we are *anatta* (no permanent self—just a cluster of vibrations), and that this anatta is an emanation of ultimate reality (the Void)—which Buddha described as "permanent, stable, imperishable, immovable, ageless, deathless, unborn, and unbecoming."† (This Void is virtually synonymous with Laszlo's Akashic quantum field.) Once we have experienced nirvana, we enter into infinite abundance and live in a perpetual state of compassion, for we see the intertwinement and impermanence of every particle in the universe—and we are moved (in the Mahayana sect) to seek nirvana for all beings. We do this by building coherent social systems—an example of how consciousness (or mental) evolution can lead automatically to physical evolution in the East. Compassion is the glue that binds together the shoreless ocean of existence and allows the formless Void to experience its own essence—to realize its destiny.

---

*Fernandez-Armesto, *Truth: A History and a Guide for the Perplexed,* 40.

†Edward Conze, *Buddhism: Its Essence and Development* (New York: Harper & Row, 1975), 40.

Taoism

One might say Taoists transmit consciousness into the world by embodying *balance* and *harmony*. Like Brahman and the Void, the *Tao* is both transcendent and immanent. "The Tao is everywhere," the sage Chuang-tzu said, "including in us." Hence, he enjoined, we must "immerse the self in the rhythm of nature and the realm of the infinite . . . becoming one with it."* The word *Tao* implies the ultimate dimension of reality. But the Tao also refers to *the way* of human life when it is perfectly aligned with ultimate reality. Like Hinduism and Buddhism, Taoism is primarily an attitude, or approach, to reality—a path of mental evolution.

Taoist metaphysics comes strikingly close to new paradigm science, wherein "energy rules matter, consciousness rules energy, and superconsciousness rules consciousness."† When we align ourselves with the rhythms of the Tao by living with grace and poise (for we are strongest and most effective when we remain supple), we tap into our higher consciousness (our inner Tao) and lead inexhaustibly abundant lives. This egoless state makes room for a higher power to enter it, enabling us to draw maximum *chi* (or vital force) from the cosmos. As a result, spontaneous action flows, combining simultaneously boundless creativity and total repose (a state known as *wu-wei.*)

When each of us is thusly flowing with purpose (without strain or abruptness, like a fish traveling downstream or Lord Siva dancing), when we have collectively brought perfect rhythm and order to all life, we create the harmony and balance needed to manifest the Tao, making it a living presence in the world. Put another way, we return the Tao to its original, transcendent state. We enable it, through our harmony and grace, to realize itself.

---

*Fernandez-Armesto, *Truth: A History and a Guide for the Perplexed,* 39–42.
†Huston Smith, *The World's Religions,* 206.

Judaism

In Judaism, the fountainhead of Semitic faiths, we transmit conscious-ness into space and time through *holiness* (that is, through the act of hallowing). *"Be holy, for I, your God, am holy"* (1 Peter 1:16). The Jews, like the Easterners, believe the essence of our nature is the same as God's:* *"God made man in His own image"* (Genesis 1:27). The Jew's task is to hallow the material world itself. By transforming matter into spirit, as it were, Jews elevate the consciousness of the world. The religious project in Semitic faiths is to make heaven and earth one by divinizing the world through righteous acts.

The material world is important in Judaism—not something to be endured, but something to be enjoyed. The Semitic traditions are more life affirming and world embracing ("God created the heavens *and the earth* . . . and beheld that it was *very good,"* Genesis 1:31) than the Eastern. The essence of the divine mission is to lean into the world and not rest until the One Truth becomes materialized through the establishment of moral order on earth.

The unidirectional linearity of time as understood by the Semitic peoples adds urgency to this mission. Hence, in Judaism, purity of intention must be balanced with positive outcomes; we must not only shift our attitude to the world, we must also change the world for the better. (Evil is real and must be vanquished—not merely reinterpreted through a lens of wisdom, as in the East.) In fact, it is through the act of changing the world (social salvation) that we change ourselves (individual salvation). We undo man's fall in the Garden of Eden not merely by erasing our forbidden knowledge of finitude, but also by making the world sacred and infinite "with labor and toil" (Genesis 3:17). We see here, as Laszlo notes, that the processes of physical and mental evolution are interwoven.

The Jews felt a passion for social justice, which would be progressively achieved, culminating in a Promised Land of peace, love, and fairness,

---

*The Jewish (Hebrew) term for God is *Yahweh*.

here on earth at the end of time. Thus, the Jews introduced the world to the concept of progress, of social and political evolution—of building increasingly complex and coherent societies within space and time until God and heaven were revealed within the flow of history itself. Since everything is a reflection of God, it follows that every inch of earth must be made holy to reflect this reality. Hence, in Judaism, an all-pervading God wishes to experience his own holiness through our hallowed actions.

Christianity

Christians transmit consciousness into the world through *love*. Jesus Christ is the most personal and intimate of the religious prophets, and Christians feel very deeply God's personal love. Christians maintain that Christ loves every one of us unconditionally, irrespective of our qualities; we love him back by loving our fellow man. (Commanded Christ: "As I have loved you, so you must love one another," John 13:34–35.) By maintaining the conviction that Christ cares for each person uniquely and without reservation and that he died on the cross for our sins, true Christians are freed of the fear and guilt that crowds out love. This conviction makes it easier for Christians, firstly, to access the Holy Spirit (Christ-nature or higher consciousness) being poured into them by the resurrected Christ, and secondly, to transmit this divine consciousness to others so it can flow freely from person to person—to assail others' egos with so much love that the shell crumbles and only the essence remains. By the end of time, the entire world is to be saturated in Christ's love.

Like its forerunner, Judaism, Christianity retains a practical and world-affirming bent; mental and physical evolution again meld into one—as love must be translated into tangible good works (in Laszlo's language, coherence-promoting actions), through which we accumulate virtue. Christians must pave the way for Christ's second coming by creating, with good deeds, a kingdom of heaven on earth. This world of love and justice is one where evolution reaches its zenith—a supercomplex, supercoherent world bathed in cosmic consciousness.

Christians often liken themselves to cells in the body of Christ. Each Christian sees herself as a vehicle for the transmission of consciousness. The early Christian cells "collectively assembled into structured communities, evolving into a new single organism" in order to further maximize their loving intelligence. This is precisely how biologist Bruce Lipton describes the evolutionary process.* Through such a process of respiritualization—of being loved back by His creation—the Christian God is able to experience His own essence.

Islam

The Muslims transmit consciousness into space and time by assuming a pose of *surrender* and *remembrance*. Not to be fooled by the mirages of the senses, the original Arab Muslims stressed remembrance of the unseen, of the eternal, above all things. *Ghaflah,* or forgetfulness of God (*Allah*), along with idol worship (deriving spiritual identity from that which is ephemeral), was the gravest of sins. The rituals and ethical injunctions of Islam (the Five Pillars of confession of faith, near-constant canonical prayer, charity, observance of Ramadan, and pilgrimage to Mecca) are designed to continually orient the devotee to an ultimate reality otherwise shrouded by the veil of the senses—and to acknowledge and bow down to its awesome majesty, power, and beauty.

The mystical branch of Islam—Sufism—goes one step further: Sufis not only acknowledge the supreme perfection of the ultimate reality, but also that they themselves harbor the essence of this divinity within them.†

Like the other Semitic faiths, Islam has a strong sociopolitical dimension. The most explicit of all religions, Islam's adherents believe that Muhammad is the final prophet and that the truth was perfected in his message. This prophet, being uniquely both a spiritual and secular

---

*Bruce Lipton, *A New Model of Evolution: Realizing Our Positive Future.* Paper contributed to the Club of Budapest's Twentieth Anniversary Conference. See www .theclubofbudapest.com.
†Orthodox Muslims would take issue with this, maintaining a sharp distinction between the Creator and His creatures.

leader, laid down highly practical and worldly instructions for his followers and placed great emphasis on social responsibility. Islam, Muslims believe, is the only religion that crystallizes the teachings of the Hebrews and Christians into specific instructions on leading a good life.

Muslims are called on to create a world that inspires remembrance and harmonizes completely with Allah's law—with the transcendent intelligence inherent in the cosmos as revealed to the Prophet. For Muslims, society must be made into a healthy organic whole; it must evolve to a higher, unified stage by their effort. This is achieved through the instrument of *jihad*—which in its purest sense means righteous exertion (but by a secondary meaning also implies holy war, when necessary, to erect a righteous empire on earth). In Islam, which is both an attitude and an ethic of worldly action, we yet again see the fusing of what Laszlo calls the evolution of consciousness (the mental domain) and of natural systems (the physical domain). By helping to create a world that conforms to Allah's will, the Muslim elevates the world, elevates his own consciousness, and perhaps allows Allah to fathom His own splendor and perfection.

Today, as we inhabit a world in which over half the globe lives under democratic regimes, we must ask ourselves to what extent are we really free? To what extent, in our daily lives, do we embody duty, compassion, harmony, holiness, love, and surrender? To what extent are we flowing with purpose? It might be argued that the behavior of the human species as a whole seems not aligned with the truth revealed by religion and now by science.

Jeremy Bentham and John Stuart Mill's formulation of the utilitarian ethos in the early nineteenth century (the ethical theory that states that the best action is one that maximizes "utility") and the Calvinists' historic obsession with proving their state of "election" (or grace) with outward symbols (perhaps to ease their anxieties about their status in the afterlife) commingled to produce a quasi-religious ethos that might be called American Capitalism. In this unique "religion," now globally practiced, one's material goods must be maximized, as they signify one's spiritual status. The size of one's bank account is an index of the size of one's soul. With the advent of the algorithm and the digital age's obsession with Big Data, our enslavement to numbers is only magnified. Reverse engineering our actions to maximize predetermined selfish outcomes is by definition incompatible with love, defined by the psychologist Erich Fromm as "commitment without guarantee."

Our present-day worldview—consumptive, short term, numbers worshipping—has infected, even hijacked, most vocations today and brought that which is most sacred under the market fold, creating a world in which right livelihood (as Buddha put it) or good work (as Jesus might say) is difficult to find. Such a worldview undermines purity of intention, and hence is incompatible with a life of spiritual abundance. Ironically, it is also incompatible with *material* abundance, as creativity flows from pure intentions.

Real flourishing, both spiritual and material, can only occur in sacred spaces that remain impervious to assault by numbers. But to the extent that we humans, being limited beings in a finite world, place value only on that which can be measured, then the market definitions of profit and success must be widened considerably if we are to remain in any way faithful to the old religions and the new science. Ideally, they would widen inexorably into that ultimate currency that emanates from our true selves: the currency of consciousness, ever augmented rather than depleted with use. And the latest technological innovations (with their distributed architecture and ubiquitous adoption) could easily be biased in favor of connection, collaboration, compassion—*coherence*—until a planetary consciousness is established. In such a world, every man could become what he truly is—flourishing from the root of his own being, unraveling his pattern of potentiality, and fulfilling his creative purpose—while helping his neighbor in this same ultimate effort.

We seem to have gone astray and lost our intuitive understanding of the spirit world, which historically has been communicated to the spiritually attuned—oracles, shamans, sages, brahmins, yogis—from the top down. Those closest to God, as it were, grasped the monistic truth of oneness through transcendent revelation and mystical intuition—through their privileged, elevated status. The truth emanated from above from an ultimate world beyond sensory perception—or even reason. Where did we take a wrong turn and lose touch with our souls? The answer perhaps traces back to Aristotle's invention of science—of inductive, or empirical, reasoning: the bold claim that universal truths could be abstracted from particular instances; that is, truth could be derived from the bottom up. Science began to take its own course (even more so after Descartes decoupled mind and matter), diverging from the fountainhead of Truth. We began increasingly to lose contact with ultimate reality—as empiricists supplanted high priests.

Fortunately, the circle has been squared, as the latest discoveries at the frontiers of science by men such as Ervin Laszlo are now empirically confirming what the ancient mystics had intuited and deeply felt so long ago. Science has again found its home in the taproot of Truth. Top down and bottom up have finally converged in a more centered metaphysics that allows us to walk down the middle path of Truth.

9

The New Answers—A Meaning for This Life and the Next

John R. Audette

For millennia, humans have engaged in a determined quest to fathom the meaning of existence. There is ample archaeological evidence from prehistoric cave art that even ancient man, though largely preoccupied with the ongoing challenge of physical survival in sometimes hostile habitats, gave serious consideration to the mystery of being. The fact that we humans have deeply contemplated this issue from nearly the beginning of our species is perhaps one of our noblest qualities and may set us apart from other members of Earth's animal kingdom.

That our early ancestors even thought to frame and consider this issue is revelatory. It speaks volumes about our formative nature, perhaps from close to the origins of human thought itself. That we are capable of deep thought of this kind signifies a great deal about our innate capacity for self-reflection, as well as our courage in facing the unknown, even when it may appear to us to be unknowable.

To ask ourselves the gateway questions *Who am I?* and *Why am I here?* is to stand at the threshold of the unknown to catch a glimpse of what new knowledge may await us in the mysterious unexplored regions. They are gateway questions because they can lead us to breakthrough discoveries and life-changing epiphanies, and because they challenge us to seek a more expansive understanding of ourselves and our world.

Almost as soon as these questions first arose in the human mind, in all probability, a concomitant expectation curiously emerged that valid answers were awaiting our discovery and could be found eventually if sought after. An enduring deep-seated expectation such as this is not merely wishful thinking, but rather may be potentially prognosticative, foretelling that valid answers would be revealed ultimately because even early human instinct compelled our progenitors to ferret them out.

Would our earliest instincts deceive us into contemplating questions that are nonsensical, or into a diligent search for answers that do not exist and can never be found? Unlikely, for we trust and rely on our instincts not to betray us. Somewhere deep in the instinctual core of our being, not only do life's big questions beg for our attention, but at the same time, many of us hear faint whispers way off in the cerebral distance that, poetically speaking, seem to come from the answers themselves.

It almost seems as if the answers are patiently waiting dormant within each individual, hoping to be discovered one day. They seem to be inviting us to engage in an engrossing process of hide, seek, and discover, leading ultimately to the highly coveted prize of heightened awareness, illumination, and enlightenment.

The great masters, saints, sages, mystics, and gurus over time have demonstrated that if we are sincere, committed, persistent, and diligent, the cosmic intelligence we seek will be revealed to us and the answers we seek will be found. Yet, determining the process we will undertake to arrive at this state of knowing and realization is paramount. It is an intensely personal process of exploration, contemplation, and deliberation, perhaps the most important self-examination of our entire lifetime.

The outcome of this self-examination provides the foundational basis of our self-identity and perception of reality. Taken together, these are the central determinants of our behavior, our actions and inactions, our choices and decisions, our values and virtues, and our basic character throughout the whole of our lifetime. This process culminates in critical life choices, such as whether we will choose to live our lives in myopic service to ourselves and our own narrowly focused ambitions, or conversely, in selfless service to others based on an abiding commitment to serving the greater good of all.

Fundamentally, we cannot answer the gateway questions addressed in this book until we first resolve the issue of death's meaning, for the meaning of death and the meaning of life are inextricably linked. Death gives life meaning and life gives death its assignment. Thus, we cannot truly know who we are or life's meaning or purpose until or unless we unravel the mystery of death. To understand life, first we must understand death and what it means to die.

Birth is our beginning, or maybe not. Death is our end, or maybe not. Surely, on the surface of things, birth and death may be what they seem, a starting point and a finish line. That could be all there is to it. If this is true, then there is no meaning to anything apart from what one experiences from cradle to grave. Experience itself, while it lasts, is the point of our being.

And, if indeed this sole, singular lifetime is all there is to it, then there is no transcendent meaning or purpose to our existence, hence no invisible overarching structure to provide any system of meaning or accountability for what we do to others, ourselves, our planet, and all creatures thereupon.

If this is the case, then all is random happenstance, accidental, arbitrary, and meaningless. What we do in life or with life matters not because there are no transcendent consequences, no implications, and no ramifications. If there are no consequences, then there is no conscience, no transcendent standard to guide or shape our conduct.

The only plausible exception could be considerations concerning the welfare of our own offspring. But in a predominately egocentric reality where individual ego-supremacy rules the day, how much of a mitigating consideration could this be? If one is guided only by the dictates of one's own welfare, then thoughts or concerns of others, including one's own offspring, would pale in comparison, taking a seat far in the back in the ranking of our priorities.

It is true that a relatively small number of human beings, the historical giants, achieved a physical immortality of sorts based on the character of the lives they have led and the weight of their enormous contributions to society. Beethoven, for example, will be remembered and celebrated for millennia after his death due to the genius and appeal of his musical compositions.

Yet, it seems doubtful that Beethoven left humanity with an amazing body of work simply to achieve this kind of social or cultural immortality. Rather, most likely he did it because it was his nature to do so and because his nature compelled him to do so. So, it appears unlikely that the prospect of achieving this kind of immortality would change one's basic character or that it would by itself motivate most humans to serve not just themselves, but the greater good.

Still, basic common sense tells us that no one life can exist in a vacuum. Life, to survive, needs other life. And to evolve and flourish, to be all it can be, life needs to engage in dynamic, reciprocal interchange with other life. Thus, "no man is an island," John Donne wisely said, and he was correct to say so because humans need each other for a host of reasons. So, then, in this context, is it true that the good of the many serves the good of the one? If so, is it also true that the good of the one serves the good of the many?

Logically, it would seem that if one statement is true, then so too is the other. Consider that society represents the many and that each individual represents the one. Social scientists for many decades have expertly documented the multiple benefits individuals derive from belonging to a social system, and this point seems inarguable. Also inarguable is the point that social systems have derived incalculable benefits due to cumulative individual contributions over time.

Thus, this symbiotic and synergistic interplay between the one and the many is axiomatic. Yet, at the same time, one cannot deny the stark reality of the greed, competitiveness, prejudice, bigotry, conflict, violence, and insularism that prevails on our planet today, as well as the destructiveness it causes both to individuals and to society.

Clearly, we do not live in a "one for all and all for one" world. Yes, there are some hopeful indicators that pockets of this thinking exist here and there, and it is most encouraging. But, there is much more evidence that "every man for himself" behavior rules the day still on planet Earth, and this could be our undoing as a species if it continues to dominate the behavior of many humans, particularly those who control most of the resources and much of the financial, military, and political global decision making impacting the vast majority of humankind.

One could argue that this is the dominant worldview of today. Evidence abounds that this is indeed the case, and understandably so given that materialism has been the prevailing worldview since the pronouncements of Isaac Newton and René Descartes. One could plausibly assert that Newtonian and Cartesian thinking is largely responsible for the nature and character of the world we live in today, in which the dominant model is "every man for himself" instead of "one for all and all for one."

Surely it cannot be otherwise, for if we truly cared about the welfare of future generations, would we have allowed Earth, "the only home we have ever known," as Carl Sagan said, to be overpopulated and overpolluted past the point of reason or sustainability? Would we continue to deny climate change and refuse to remedy it? Would we continue to behave in an irresponsible and reckless manner by arrogantly persisting in nuclear proliferation and the production of weapons of mass destruction? Would we continue the accumulation of unprecedented global economic sovereign debt that has effectively bankrupted several successive future generations in America, Europe, Japan, and elsewhere, which some call "weapons of mass financial destruction"? Would we continue to turn a blind eye and a cloth ear to over 50 percent of the population currently suffering in starvation and penurious squalor? Methinks not, and methinks that these things evidence a pervasive lack of caring, compassion, and altruism on our planet today, which if left without remedy will likely be the cause of our undoing.

While most human beings ask themselves life's big questions at one time or another during their earthly existence, it is probably true that only a small percentage of human beings actually become sufficiently motivated by them to seek reliable answers in a steadfast manner. Only a certain inspired fraction of the population is likely to dwell on these issues and become preoccupied with the quest for satisfactory answers. Fewer still make a commitment to walk a steady path toward enlightenment, along "the razor's edge," in the words of Somerset Maugham.

If there was ever a time when everyone on the planet should be contemplating these gateway questions and finding reasoned, reliable answers to them, it is now while we still have time, before we self-destruct. We live in deeply troubled, perilous times, with our very survival in jeopardy. There is but one cause underlying all this turmoil. It is none other than flawed thinking borne of erroneous perceptions of reality.

If one day the erroneous perceptions of reality and the flawed thinking they spawn turn out to be the root cause of our species' extinction, then our own stubborn ignorance will be to blame, for solid scientific evidence exists that should motivate us to completely reevaluate who we are, why we are here, and what our true relationship is to one another and all living things.

Stubborn ignorance is a petulant refusal to open one's mind and change one's views and one's behavior in the face of certain incontrovertible facts, even when credible knowledge and evidence are available. It constituents stupidity of the worst kind.

That many of us refuse to closely examine scientific evidence concerning the gateway questions and contemplate the meaning of them is sad and tragic, for it imperils our future in no uncertain terms. This potentially fatal penchant of ours to ignore the "truth that will set us free" does indeed deserve intense scrutiny and analysis. It would serve us well to closely examine our own recalcitrant refusal to evolve by embracing the credible knowledge available to us from the science of consciousness. Toward that goal, the following theoretical framework is postulated for consideration in the hope that it may provide a better understanding of this most lamentable human proclivity.

Five Personality Types in Humanity's Search for Meaning

To stimulate contemplation of this weighty cornerstone issue, I hypothesize that there are five main personality types that together may describe how and why most humans manage their consideration and pursuit of knowledge concerning life's gateway questions, or in the main perhaps, their refusal to do so.

The first personality type is termed the "Materialist Personality." Atheists, nihilists, existentialists, and many conventional, traditional, and mainstream scientists belong to this category, which maintains that physical matter, as well as information and energy in the quantum realm, is all there is to reality. There is no larger unseen transcendent spiritual reality or ground of being from which it all emerges. Those with this personality belong to the lineage of Newton and Descartes.

Materialists contend there is no divine realm and no afterlife and no creator. We are merely physiological/biological creatures on a planet orbiting a star in a random cosmos filled with billions of stars, fated to live out a relatively brief existence, destined for nothing but oblivion. The sum total of our being, our identity, our consciousness, is the physical body and nothing more. When it no longer exists, we no longer exist. The universe and everything in it, including us, is one huge machine of pure happenstance.

The Materialists assert that this lifetime is all there is for us as individuals. There was nothing before it in terms of discrete individual awareness and nothing that comes after it except nonbeing. When you're born, you grow into awareness that is purely the result of normal brain functioning and sensory perception. When you die, your awareness and consciousness cease because your brain is no longer operative. You exist no more and no point was served by your existence except whatever temporal experience you encountered during your brief lifetime.

We humans are here, Materialists contend, simply to experience whatever this one single, solitary lifetime brings to us. They feel the sole purpose of life is to reproduce and also to amass the largest estate possible to leave behind for heirs as their inheritance, preferably our offspring, who will perpetuate our bloodline and gene pool, or better still, male offspring who will also perpetuate the family name and crest.

Materialists firmly believe that they have correctly answered life's gateway questions in this manner. They are largely content that their perspective is valid. No amount of evidence to the contrary, no matter how powerful, will convince them otherwise. As such, they have managed to confine life's big questions to this point of view and do not feel the need to ponder or pursue them further, damn all evidence to the contrary.

Materialists would assert that our basic instincts guide us and command us, sculpted over millennia during our physical evolution on this planet, starting out as protozoa. Our primary imperative, they argue, is to survive, control, and dominate our environment and all those around us. Our will to live and to be dominant over all things is our strongest urge. We are compelled by our inherent nature to overcome threats and challenges, to procreate, and to stave off death for as long as possible, for this lifetime is all there is and to believe otherwise is to succumb to pure fantasy.

Materialists do not question why we exist, or if we should exist, or what the purpose of existence may be. In their minds, we simply exist and that's all there is to it. We are not biologically directed to question it or debate its meaning. Rather, our prime directive is our will to survive and procreate and dominate all others as the fittest in the competition for survival to make our physical existence endure as long as possible and be of the highest possible quality. For when it's over, it's over. End of story. Nothingness. Nonexistence.

The second personality type is called the "Nonchalant Personality," which is the "Avoider" personality, one that makes little or no effort to tackle life's big questions. This is a lackadaisical, superficial, and casual posture, even dismissive, in that one devotes little or no time to the matter. Agnostics, I submit, belong to this personality type, insofar as they resign themselves to the neutral position of declaring that they simply do not know and cannot know the answers to life's big questions. Some may go so far as to claim that such matters are unknowable and therefore off limits. They would label these issues to be the great imponderables.

The Nonchalant Personality chooses to go through life ignoring or avoiding the big questions as best they can, opting to comfortably pretend that these issues will magically sort themselves out over the course of time with no need for inner work or preparation. Their rationale is essentially one of fatalism, as if to say, it is what it is and it is whatever it will be, so no need to be concerned about it during one's lifetime. Better to brush it all aside.

The Nonchalant Personality can be shallow and does not feel adequately equipped to find or formulate answers. Instead, they choose to go through life in an escapist fashion in relation to the gateway questions, preoccupied with all manner of convenient dramas, distractions, and detours. They are not motivated to critically examine the vast body of evidence related to these issues, except perhaps in a cursory manner, giving little or no merit to the accumulated repository of knowledge on this subject.

The Nonchalant strategy of denial can work very well, that is until the fortress of one's self-identity begins to disintegrate on one's deathbed. When the support pillars of one's self-concept begin to crumble and the walls that bound self-identity begin to cave in, one is left with no good inner place to go except severe mental meltdown caused by an identity crisis of maximum extremes.

Thus, as one's body and one's possessions begin to fade away concomitant with one's approaching physical death, the Nonchalant approach reveals itself to be wholly unsatisfying, unrewarding, and unfulfilling, leaving one feeling desperate, alone, frightened, and empty in the end. This begins the ugly confrontation with having lived a lie, with the failure to become spiritually awakened during one's lifetime. This same syndrome often applies to the Materialist personality type as well.

The third personality type is the "Abdicator Personality." This personality type was aptly described by C. S. Lewis, popularly known as the author of The Chronicles of Narnia series. In a style befitting his prodigious intellect and creativity, he offers in his incisive work titled *The Abolition of Man* a candid discussion about "the Conditioners," who classically enable the Abdicator Personality type.

The Conditioners, according to Lewis, *"know how to produce conscience and decide what kind of conscience they will produce . . . they are the motivators, the creators of motives."* In today's lexicon, Conditioners would be considered thought leaders, opinion makers, trendsetters, style makers, those programming our views, attitudes, preferences, values, priorities, beliefs, prejudices, practices, and more.

Individuals with the Abdicator personality type abdicate their decision making concerning life's big questions to others, the Conditioners. They surrender their personal power, opting instead to substitute intellectual rigor and analysis for blind faith in someone or something else, the Conditioners, whether deserving or undeserving of their confidence and trust (often the latter). Conditioners can come in the form of government or other social institutions, political ideology, religion, spiritual leaders, a charismatic demagogue, a ritual, modern media, money and materialism (worship of the "Golden Calf"), or some other external distraction or alternative.

Abdicators are indoctrinated by their Conditioners, accepting whatever information they sanction, even if it is false. Indoctrination suppresses individuality and calcifies independent thinking. Codified dogma stifles critical thinking and makes one rigid in attitudes and perceptions. One's life, therefore, becomes an extension of the Conditioner's programming, which, in the end, can turn out to be entirely false, leaving the Abdicator disillusioned, disenchanted, and crestfallen. This begins the painful consequence of having lived a lie.

Abdicators willfully abandon their innate capacity for critical thinking, submitting to the alluring and abiding influence of their chosen Conditioners. Abdicators sublimate their independent judgment and replace it with full reliance on the teachings, literature, theology, and counsel of their Conditioners to whom or to which they have succumbed. They are content to remain under the umbrella of total belief and faith in their Conditioners of choice, confident that the answers provided by them are completely trustworthy and valid.

Abdicators decline to engage in serious independent pursuit of knowledge or direct experience concerning life's big questions, finding it far more convenient and expedient to rely solely on their Conditioners for all of the answers and information needed. Many Abdicators inherit this orientation from their parents, passed down from generation to generation. Their perspective is often purely the result of their upbringing, socialization, and cultural conditioning.

To be sure, Conditioners have their own karma to sort out for the damage they cause, but so too do Abdicators. Being an Abdicator does not relieve an individual, in any measure, of the personal responsibility each and every human being has, given their innate capacities and capabilities for critical thinking and independent acquisition of knowledge, to seek out and find his or her own reasoned and reliable truth to life's big questions.

The fourth personality type is the "Experiential Personality." This personality type, to be admired if not emulated, generally springs from powerful direct spiritual experiences of one kind or another that provide firsthand knowledge about life's big questions. This could occur from one or more of any number of causal factors, typically referred to as epiphanies, mystical experiences, peak experiences, exceptional experiences, and the like.

These include spiritually transformative experiences such as near-death experiences, after-death communications, nearing-death awareness, past-life recall, out-of-body experiences, and more. Additionally, they include nonlocal consciousness experiences such as spontaneous healing, telepathy, precognition, clairvoyance, mediumistic communications, and many more. Also, they include experiences derived from disciplined spiritual practices such as meditation, prayer, fasting, chanting, rhythmic repetitive movement (tribal dancing), sensory deprivation, self-hypnosis, extraterrestrial encounters, angel encounters, psychedelic drug ingestion, transcendent sexuality, and much more.

In these extraordinary situations, individuals enter into an altered state of heightened consciousness in a transcendent state, generally apart from and outside of the physical body, sometimes while in a state of clinical death, in which they often enter another realm or dimension of reality where they encounter divine beings, deceased loved ones, friends, pets, and more. It is typically a profoundly life-changing experience, which commonly inspires "experiencers" to undergo major shifts in personality, values, attitudes, and behavior. It often brings about dramatic shifts in one's perception of and perspective on reality, both the life now and the life after.

Individuals with this personality type are blessed with glimpses of the other side, that is to say, the transcendent dimensions or realms. They no longer fear death. They no longer wonder about life's big questions. They discovered the answers for themselves during the life-changing experiences they had. They are now in a place of knowing, of certainty, based on their own direct personal experiences. Socrates, I believe, was an Experientialist, which is why he willingly accepted his prosecutor's call for the death penalty.

The fifth personality type is called the "Empiricist Personality." This personality type is objectively and dispassionately data-driven and evidence-based in their quest for answers. They are interested only in facts based on irrefutable and incontrovertible evidence. They seek only reasoned and reliable truth, not opinions about it or distortions based on it. They have no time to waste chasing illusions, falsehoods, fantasies, or half-truths. Rather, they embrace timeless incontrovertible wisdom and certain knowledge, as well as indisputable facts.

An Empiricist believes one should formulate answers to the big questions based on objective scrutiny of all available evidence. Views, opinions, and beliefs should all be grounded and based on data—on knowledge—deemed to be reliable and credible by the standards of frontier science.

Empiricism has its limits of course. It is not a perfect way of knowing. Data can be flawed and interpretations of data can be faulty. But unlike a dogmatic mind, the empirical mind will change with the discovery of new knowledge. As new data emerges, and new interpretations along with it, then so too does the Empiricist change his views, opinions, perspectives, and perceptions, correspondingly and proportionately.

If the Materialists, the Nonchalants, and the Abdicators, as well as their Conditioners, could view reality from the vantage point of someone with an empirical perspective, I opine that their worldviews would undergo dramatic transformation into that of the Empiricist Personality. I speak from experience here because I underwent this same metamorphosis myself. It was eye-opening and life-changing for me.

I began my work in this field from the orientation of an agnostic at the age of twenty-two. I had no preconceived notions or opinions about anything in relation to life's big questions. But, having done the heavy lifting and the hard work laboring in the fields of accumulated knowledge from the science of consciousness, now there is no doubt in my mind that we are indeed eternal spiritual beings whose purpose during physical existence is to learn to love all things unconditionally, including ourselves. I also learned that we are all one, coming from and returning to the same Source. It is now abundantly clear to me that what we do to others, we do to ourselves, and that we should embrace a "one for all and all for one" orientation.

But I caution once more, do not take my word for it, and do not accept my example as one that should relieve you, the reader, of the joy of discovering your own truth. If your process is truly that of a high-integrity, fully committed, open-minded Empiricist, I have full confidence you will arrive at the same destination I did, or one very similar. If not, then we should most certainly compare notes one day. There may be something we can learn from each other.

As Socrates said at his trial, *"I know that I know nothing."* I agree wholeheartedly, for the more I learn, the less I know, to reference an old adage. Yet, based on the credible evidence before us now, I am confident that Eternea's Seven Statements and its Fifteen Elaborations, which can be viewed at Eternea.org, are indeed credible and accurate answers to life's gateway questions, at least for the time being, based on what we know at the present time. This is subject to change of course as our knowledge expands. I am open to that possibility, though extremely doubtful it will ever come to pass given the powerful convergence model on which Eternea bases its assertions regarding the nature of existence.

I urge you to make haste on your own path to finding truth because it is plain to see that the materialist model is stealing or certainly compromising humanity's future, conditioning many humans to engage in obsessive consumption and mindless exploitation of Earth's precious resources, deceiving them into falsely thinking that this lifetime is all there is, so we better get while the getting is good. Such thinking is leading us to galloping ruin.

The sooner we abandon materialism as flawed and invalid, the faster we can co-create a truly magnificent future for Earth and all its inhabitants. This wonderful potential destiny will escape us, however, until or unless we discover our true spiritual identity and our role in the larger reality in which all things are truly one, interdependent and interconnected, coming from and returning to the same Source.

In honor of all the brave souls from ancient history like Socrates who martyred themselves in the name of truth, as well as those from more recent history like Gandhi, it is past time for us all to end the incessant debate about whether consciousness survives bodily death.

Instead, we should turn our energy and attention promptly to the more important question of what it means that individuated consciousness is indeed indestructible and eternal.

For what it's worth, I assure you that the discovery will come into your consciousness eventually that you are not your current body, or your current social identity. This discovery will happen, sooner or later, at the very latest just before the moment of your own death. Your body and your social and ego identity, your net worth and material possessions will take their leave and abandon you at the end of your days here on Earth. Best to realize now that they are not the sum total of who you are really, not even in a tiny fractional sense. You are none of that really, but yet you are so much more.

One day, we all will discover that this lifetime and all of its trappings are not all there is to reality, not even close. Rather, it's only a momentary blip on the screen of eternity. Our essence—our spirit—our soul—our consciousness is everlasting, moving continually in and out of different physical forms in different times and places. I assert that the search for credible answers to life's big questions, if honest, is likely to lead us to this inescapable truth about the life now and the life after.

The eternal now and the infinite here are ours to experience in the present moment, one present moment to the next. We can make of it what we will, heaven or hell. But we are accountable for what we do, not only to our Self, but rather to our larger, greater, higher Self, the divine Self that knows that all things are one, derived from the same Source, the Source that makes all of creation possible and endows it with the tremendous gifts of abundant resources, free will, and the capacity to evolve, ultimately, into perfect expressions of unconditional love.

Upon release of our eternal soul from our current temporal bodies, we will return to Source to experience the oneness of all things and the profound beauty of unfathomable unconditional love originating from Source, which we will then discover to be the organizing principle of the universe. It is then that we will know and fully understand that what we do to others truly we do to ourselves, for despite the illusion of materialism and the lie of duality, all things are interconnected in the magnificent matrix of creation.

If and when the day comes that these insights are generally accepted by a critical mass of humanity, it could inspire sweeping changes in basic human nature and, hence, in the nature of social, political, economic, and religious institutions. It could engender a shift from the current "every man for himself" model of social organization to the more enlightened model of "all for one and one for all," which is the essence of the Eternea Strategy for individual and social change, as well as the New Paradigm advocated by Eternea's chairman, Ervin Laszlo.

10

The New Answers and the Challenge

Reconciling Science and Spirituality with Life

Garry Jacobs

Life is the meeting place between the objective external world in which we live and the subjective inner world of consciousness with which we experience it and by which we act on it. Therefore, life is the ultimate litmus test for knowledge. In spite of the remarkable, unprecedented achievements of physical science and technology, we still find ourselves groping blindly for solutions to the most basic problems of life—peace, political stability, economic security, social harmony, cultural relations, psychological fulfillment, and spiritual purpose. Unless and until we learn how to master these problems, all our scientific knowledge and technological mastery of the external material world in which we live—the subatomic microcosm and the intergalactic macrocosm—are inadequate. So too, mind's quest for spiritual knowledge of inner and higher states of consciousness is incomplete unless it teaches us how to live fully in peace, harmony, and mutuality with our own selves and other people. Science urges us to seek that knowledge in the future by more rapid technological advancement. Spiritual traditions counsel us to rediscover the significance of ancient wisdom. Science and spirituality grope from two different poles of existence toward the same goal, and very likely the fulfillment of each lies in the meeting point between the material and spiritual planes of existence that we know as Life. The knowledge humanity needs most is that which will provide complete, fulfilled, and ever-increasing mastery and self-mastery of both the inner and the outer domains of our existence. Knowledge is power and true knowledge is power for the perfection of ourselves and for the world we live in.

The New Answers and the New Paradigm

Ervin Laszlo's essential insights, based on the new paradigm he is enunciating, respond to fundamental questions about the nature of reality and the destiny of human beings. Do we live in a meaningless universe governed by random chance and blind necessity, or is there purposefulness in nature? Does the evolution of the physical universe, life-forms, and consciousness have a direction and goal, or is it simply the result of an infinite number of infinitesimal accidents? If chance governs all, how can we explain the remarkable precision, orderliness, and purposefulness found in physical forms, living species, and conscious mentality? If the ultimate nature of reality is inanimate, inconscient material substance, how can we explain the emergence of animate, subconscious life-forms and conscious mental self-awareness? If evolution results in the emergence of life and consciousness from dumb material substance, then what does it tell us about the fundamental nature of matter in which life and consciousness are in some sense involved and potential?

These are questions that in various forms have perplexed humanity and been debated without resolution for millennia. But no matter how difficult to answer in terms satisfying to science, religion, and personal experience, their profound relevance to humanity justifies every attempt at reconciliation. For our answer to these questions determines the meaning of our existence, the validity of our quest for knowledge and truth, the value of goodness, love, and everything else we cherish as human beings.

Laszlo's answer integrates insights from science, philosophy, and spiritual experience. Throughout history these three ways of knowing have been the primary means by which humanity has sought knowledge of the unknown, complemented by the inspirations of poets, musicians, and artists giving expression to inexpressible mysteries of existence. In different periods and places each of these powers of consciousness has assumed a dominant influence over the others and often leveraged its preeminent position to discredit, suppress, or reject alternative ways of approaching reality. Today we live in a schizophrenic age in which unprecedented scientific knowledge and technological mastery of the external, physical world overshadow and threaten to efface fundamental insights regarding the deepest and highest reaches of human experience.

Our self-ignorance seems to grow in proportion to the growth of our scientific knowledge. The deeper we delve into the mysteries of material nature, the more we become aware of the inadequacy of objective, physical principles and processes to explain the organic unity of individual consciousness and human experience. Everywhere we find evidence of an increasing sense of alienation and loss of harmony with ourselves, other people, and nature. An unrivalled sense of power and growing sense of impotence coexist side by side. A balanced attempt to reconcile and integrate diverse viewpoints of reality is of greatest importance to heal the gaping holes in our self-knowledge and relationship with the world around us. Efforts of this kind deserve the same respect as that accorded to the efforts of modern science to perceive hidden patterns and relationships in the material world.

As a member of an international think tank of scientists, artists, diplomats, and civil society leaders, I am continuously involved in efforts to understand and reconcile different ways of knowing. I am often surprised by the facility with which the most extraordinary human experiences are logically reduced to chemical and electrical events by idealistic, deeply feeling individuals, unaware of the inherent contradiction between the materialist hypothesis and the affirmation of nonmaterial values such as freedom, love, dignity, and truth. For if, indeed, we are nothing but complex formations of material substance, there can be no ultimate significance to our lives, the truths we affirm, or the values we cherish. I am equally surprised by the frequency and facility with which distinguished physicists and biologists affirm the ultimate importance of intangible spiritual values and ideals far beyond the purview of physical science. *Science and spiritual experience remain two real, but separate, domains in search of a synthesis.* Confronted with such a diversity of perspective, what I write here necessarily represents my own personal views and effort at reconciliation.

Having said that, this book affirms meaning in the patterns of cosmic evolution, and Laszlo discovers a movement of superconvergence between the cosmic macrocosm and the microcosm of conscious human experience. He begins his essay by pointing to a fundamental truth of existence that is confirmed by both science and spirituality. As modern physics discovered a century ago and the Vedantic sages realized thousands of years earlier through personal experience, all the forms and objects that constitute the material world resolve themselves at a more fundamental level into forms of energy, vibrations of a more ethereal, less material reality. During the twentieth century, science progressed very far in deciphering the forces that constitute material forms and govern their interconversion. It has rejected the evidence of the physical senses in favor of the mathematical precision of abstract equations. Yet it clings to the unproven and unprovable hypothesis that this fundamental energy is the sole and ultimate basis of existence.

Limits to Rationality

Laszlo effectively challenges this position. He argues convincingly that the available evidence, even the evidence of physical science, is not adequately explained by a purely material hypothesis regarding the nature of reality. While he may not provide all the answers, he certainly asks the most fundamental questions. He urges us to consider whether there is an alternative hypothesis that is equally or more logical and consistent with the facts. He invites us to discard the prevailing conventional wisdom in favor of a truly rational exploration of possibilities, while at the same time reminding us that rationality has its limits. At the lower end of the spectrum, psychology, sociology, and other social sciences confirm that we are not, or not yet, truly rational beings. Our attempts at rationality are invariably accompanied by heuristic errors, social constructions of knowledge, personal preference, prejudice, superstition, and the limitations of our own cultural and egoistic perspectives. Nor is science an exception. It too rests on a social construction of reality, rigorously enforced. Whatever our capacity for reason, both history and contemporary events testify to our continued insistence on beliefs and behaviors inconsistent with rationality. We have largely liberated ourselves from unquestioned faith in the sanctity of religious dogma, only to resurrect it in new form as the sanctity of scientific theory.

At the higher end of the rationality spectrum, great scientists unanimously testify to the fact that rationality itself has its limits. Evidence abounds that the greatest scientific discoveries have been the result of intuitive rather than linear, rational mental processes.[*] Even mathematics, the queen of rational sciences, is founded on insight and intuition. "What then is mathematics if it is not a unique, rigorous, logical structure? It is a series of great intuitions carefully sifted, and organized by the logic men are willing and able to apply at any time."[†]

The scientific method has proven highly effective for falsifying hypotheses, but very rarely has it been the source of the new knowledge

---

[*]Karl Popper, *The Logic of Scientific Discovery* (New York: Basic Books, Inc., 1959); Richard Courant, preface in R. Courant and David Hilbert, *Methods of Mathematical Physics* (New York: Wiley, 1937, 1989), v; and Gerald James Holton, *The Advancement of Science, and Its Burdens* (Cambridge, Mass.: Harvard University Press, 1998).

[†]Morris Kline, *Mathematics: The Loss of Certainty* (New York: Oxford University Press USA, reprint edition 1983), 312.

that it is used to validate. Insight and intuition have been the principal sources of knowledge, both scientific and spiritual. They usually arise only after the mind exhausts its capacity for rational analysis and falls silent. Their processes remain beyond explanation by reason and validation by the scientific method.

These limits to rationality at both ends of the spectrum present a perplexing dilemma to anyone seeking to explore knowledge beyond the reach of physical science. To those who cling to the sense of certainty and security offered by contemporary scientific thought, these limits pose a disconcerting threat. Those who understand the limitations of mental cognition regard it as good counsel for open-mindedness, tolerance, and humility.

Unresolved Issues

Laszlo generously recognizes the greatest achievements of modern science, but draws on their evidence to frame some alternative conclusions that run counter to the trend of contemporary, materialist thought. He points to the three great unresolved issues of science: the origin of the universe or what existed before the Big Bang, the origin of unicellular life, and the nature of mind and consciousness. Science seeks to explain all three as extraordinary events—singularities—that are so statistically improbable that they may have occurred only once in the entire history of our universe. He goes further by citing the extraordinary improbability governing the cosmic constants that are essential to support life in the universe. Here he does a great service by exposing the points at which science morphs into speculative philosophy without openly acknowledging the transition. The anthropomorphic hypothesis that humanity happens to live in a universe that is suitable for life simply because it is the only one in which human life could thrive is pure philosophical speculation disguised as theory. Leading physicists have affirmed that the same is true of many of the postulates of string theory.*

---

*Lee Smolin, *The Trouble with Physics* (Boston: Houghton Mifflin, 2006), xiv.

The problem of life is equally perplexing. For even if simple organic molecules can form spontaneously from their constituent elements and join together under certain physical conditions, that does not prove the highly improbable hypothesis that life actually arose in this fashion. The materialist postulate appears plausible only so long as we refuse to consider alternative explanations. It requires a huge leap of speculative faith to assert that the random self-assembly of increasingly complex molecules is sufficient to explain the emergence of increasingly complex life-forms exhibiting characteristics of self-organization and self-replication according to principles that run counter to the laws of entropy governing purely material phenomena. Life progresses from lower to higher levels of order and organization, while matter moves in the other direction. The fact that we accept these premises as not only logical but plausible only reflects the degree to which the materialist hypothesis has been implicitly accepted as the sole premise on which a rational view of reality can be fashioned.

But it is when it comes to the problem of consciousness that the material hypothesis is found most wanting. The case supporting the equivalence of mental consciousness with the activities of the physical brain is far from convincing. It may be true that our conscious functions are related to centers and activities in the brain, but that does not prove causality. Even if compelling evidence were to exist with respect to basic sensory and memory functions similar to those now exhibited by computers, it requires a huge leap of speculative imagination to reduce the highest, most rarefied experiences of consciousness—the quest for truth, the thirst for immortality, the insatiable joy of freedom, and the intuition of perfect love—to purely physical processes, no matter how complex the network of neurons or the regulatory feedback mechanisms. *To imagine that clay and sand have given rise to conscious, feeling, aspiring human beings over so many millions of years is an act of speculative philosophy, not scientific fact.* Once again, it is only the unwillingness to examine alternative hypotheses that compels us to work on this untenable and unprovable premise.

Contemporary science has sought to address this paucity of evidence by terming the evolution of animate, subconscious life as an emergent property of inanimate, insensible, inconscient matter and the evolution of self-conscious mentality as an emergent property of subconscious life. But coining a scientific term to describe something inexplicable and unprovable by present knowledge cannot pass for a rational explanation. *Terms such as emergence and self-organization are merely descriptions, not explanations.*

Atheism, agnosticism, and materialism have performed a great service over the past few centuries in cleansing spiritual knowledge from dogma and superstition. Similarly, pointing out the speculations, inconsistencies, and contradictions implicit in many prevailing scientific viewpoints is an equally great service to the progress of knowledge. It is not necessary for us to conceive or accept an alternative hypothesis before we concede that current premises are purely speculative. What is needed is an open-minded, relentless inquiry and search for more plausible explanations.

Alternative Hypotheses

Laszlo argues that these unresolved issues justify a reevaluation of the fundamental premises on which modern science is based. As long as the founding premises are unproven, is it not rational and even necessary that we explore the alternatives? Here he draws on philosophy and spiritual experience to put forth an alternative hypothesis. His premise is similar to that of the Vedanta, that the fundamental nature of reality is a field of pure consciousness. Modern science has discovered that material forms are actually vibrations of a more subtle and invisible energy that our senses perceive as solid, physical forms. Laszlo argues that both the material forms and the energy of which they are constituted are themselves perceptible, measurable vibrations of the more subtle and universal Akashic field, informed by consciousness. Consciousness manifests as sensation, cognition, and self-awareness that emerge in the progressive evolution of the lower and the higher forms of life.

From the vantage point of materialism, this explanation appears far-fetched for the simple reason that it cannot be confirmed by the physical methods applied by modern science. But that only leads to an obvious conclusion that was fully apparent to the thinkers and scientists of the Enlightenment. Modern experimental science was only conceived and developed as a method applicable to external material objects and forces. Its pioneers never imagined that this methodology would be applied in an attempt to explain the nature of life, mind, and spiritual experience.

Evidence for the New Answers to the Nature of Consciousness

The new answer to the nature of consciousness query is equally rational and consistent with all the known observations of modern science. In fact, more so. Because it does not depend on an extraordinary series of singularities to explain the physical, biological, and mental evolution of matter, life, and mind. For if the fundamental reality is pure consciousness somehow involved and concealed in material form, then its progressive manifestation and expression through higher and higher forms would be natural and inevitable. Already science acknowledges that the most solid, stable material forms are actually constituted of physically detectable energy vibrations. It is only one more step to concede that the energy vibrations we detect physically may in turn be manifestations of a more subtle reality that has the property of consciousness. In that case the inherent orderliness, organization, and purposefulness we find everywhere in Nature are readily explained. In that case, life would be a manifestation of this consciousness building up biological forms just the way physical energy builds up material forms. The gradations of mental experience we find in higher animals and human beings would represent the progressive manifestation of this inherent consciousness. The view of brain as a physical receiver through which mental consciousness expresses, rather than a generator of consciousness, is equally consistent with observable scientific facts that find a correlation between brain function and human perception, emotion, and consciousness.

This viewpoint is equally or more logically consistent and plausible than the materialist hypothesis. Moreover, it is supported by the testimony of our own self-experience as human beings. For no matter what science may postulate, human beings intuitively experience their own thoughts, feelings, emotions, aspirations, and values as real in themselves and not merely expressions of physiological processes. As Schrödinger put it, "The scientific picture of the real world around me is very deficient. It gives a lot of factual information . . . [but] it cannot tell us a word about red and blue, bitter and sweet, physical pain and physical delight; it knows nothing of beautiful and ugly, good or bad, God and eternity. So, in brief, we do not belong to this material world that science constructs for us . . . the scientific worldview contains of itself . . . not a word about our own ultimate scope or destination."*

Scientific evidence, logical consistency, and subjective personal experience do not constitute proof. They only present a compelling justification for examining all the available evidence through appropriate means. Therefore, Laszlo draws on the testimony of spiritual traditions around the world over thousands of years in support of this hypothesis. He cites also recent testimony of those who have undergone near-death experiences. Although his approach to these controversial issues is interesting, it is unlikely to be convincing or accepted as adequate evidence by those who have not had direct experiences of this type. But it should be sufficiently substantial and consistent to demand an open-minded consideration rather than a cynical rejection.

Objectivity and Subjectivity

Resolution of the conflict between these two contradictory views of reality is rooted in the fact that all objective knowledge is the result of subjective processes of cognition. Science studies the objective field

---

*Erwin Schrödinger, *Nature and the Greeks* (Cambridge: University Press, 1954), 93.

of external reality. Spirituality studies the subjective field of psychological self-experience. But all knowledgeable, even scientific knowledge of external material phenomena involves a subjective process of perception, mental interpretation, social and psychological construction of reality. "Our perception of reality is largely modelled from beliefs and assumptions that are typical of the society and culture to which we belong. What we know, what we consider true and right, the behaviors we adopt, are all influenced by the social/cultural environment in which we live. This process happens through the internalization of a 'reality' that occurs during the socialization process."*

The insistence on pursuing a purely materialist explanation for life and consciousness is a consequence of the phenomenal success of early science in discovering the processes of material nature. A long, wandering detour over several centuries from the dawn of the Enlightenment to the present day has led us to deny the essence of our own most intimately human experiences. In their first turn away from the sanctity of religious dogma, the thinkers of the Enlightenment sought for an external, objective means to determine truths about the external material world in which they lived. They relied on acute observation, repetitive verification, measurement, and mathematics as instruments well suited for the study of objective physical phenomena. They sought to eliminate the intrusion of corrupting influences such as personal preference, prejudice, religious belief, and prevailing social conceptions. As a result, they developed an impartial, impersonal objective scientific method that proved highly effective for the study of external material objects. The method was objective in the sense that it dealt with objects and related phenomena that could be observed and measured through objective means externally. Enlightenment thinkers such as Newton and Descartes did not believe or assume that all aspects of reality could be studied through the scientific method or ultimately be reduced to

---

*Alberto Zucconi, "The Need for Person-Centered Education," *Cadmus* 3, no. 1 (2016): 1–26.

a purely material basis. In devising a method to minimize the intrusion of personal preference, they never intended to deny the existence or validity of subjective dimensions of reality and self-experience or to assert that these nonmaterial realms could be adequately studied and explained in purely physical terms.

In later centuries, the very notion of objectivity underwent a subtle shift in meaning. It began as the study of the external world of objects, because that alone lent itself to collective study by objective means, and gradually morphed into the assertion that only the objective, external world is fully real and all else is subjective interpretation or distortion of reality. As Canadian mathematician William Byers explains it, "The word 'objective' has two distinct meanings: independent of personal opinions or impartial; and independent of mind. Science confuses these two meanings."* The essence of science is the pursuit of knowledge that is free from prejudice and personal bias, but knowledge can never be free from subjectivity. There is no knowledge of the objective world that does not involve subjective processes of consciousness. *All knowledge is subjective.* Instead of confining its meaning to the study of material nature by impartial means, positivism asserted that only that which can be observed materially and studied as external object is real. All else is not merely subject to preference and prejudice but nonexistent or merely a derivation from material phenomena. Such a premise may be valid in philosophy, for philosophy admits of an unlimited number of diverse viewpoints. But it has no rational scientific validity.

The Evolution of Consciousness

Laszlo's thesis is that the universe is a progressive manifestation of consciousness that is evolving toward supercoherence. Evolution is a double movement. Externally, it expresses in the evolution of physical

---

*William Byers, *The Blind Spot: Science and the Crisis of Uncertainty* (Princeton, N.J.: Princeton University Press, 2011), 103–104.

and biological forms studied by science. Internally, the evolution of these forms is driven by and expresses an evolution of consciousness that seeks higher forms through which to more fully express its inherent powers.

The evolution of coherent systems and the evolution of consciousness are complementary aspects of a unitary evolutionary process. Physics charts the physical evolution from infinitesimal atoms and molecules to the solar systems, galaxies, and other bodies that populate the known universe. Biology charts the biological evolution from infinitesimal, unicellular life-forms to the most complex, adaptive human species. We also see clear evidence of the evolutionary progression of mental capacities from the subconscient sensitivity and responsiveness of plants to the instinctive behaviors prevalent in the animal kingdom to the mentally conscious and self-conscious awareness of human beings, with our capacity for complex language, abstract thinking, and self-reflection. Cosmic evolution proceeds toward higher levels of organization, complexity, adaptability, and knowledge of self and the world. The expressions of consciousness observed in life-forms do not necessarily reveal the characteristics of consciousness itself, only its power for manifestation. If that manifestation is progressive, there is no reason to think that it has reached its natural limits or that life and mind as we know them are its highest term. For both life and mind are obviously limited, imperfect phenomena that are still in the process of evolving.

If sentient animate life can emerge out of matter and conscious mentality can emerge out of subconscious life, it is reasonable to postulate that higher forms of consciousness may yet evolve out of our mental consciousness. The ultimate nature of consciousness may differ as much from our present mental experience as our mental experience differs from the awareness of lower animals and plants. Supercoherence may be another term for a perfect manifestation of the full powers of that consciousness. This possibility is confirmed by the phenomenon of genius, the testimony of great scientists and artists, and the massive historical record of spiritual experience, as described below by Sri Aurobindo.

We speak of the evolution of Life in Matter, the evolution of Mind in Matter; but evolution is a word which merely states the phenomenon without explaining it. For there seems to be no reason why Life should evolve out of material elements or Mind out of living form, unless we accept the Vedantic solution that Life is already involved in Matter and Mind in Life, because in essence Matter is a form of veiled Life, Life a form of veiled Consciousness. And then there seems to be little objection to a further step in the series and the admission that mental consciousness may itself be only a form and veil of higher states which are beyond Mind. In that case, the unconquerable impulse of man toward God, Light, Bliss, Freedom, Immortality presents itself in its right place in the chain as simply the imperative impulse by which Nature is seeking to evolve beyond Mind . . . *

This hypothesis is too consistent with the observable facts and our own subjective self-experience to be rejected simply because it cannot be proven through the methods of physical science. Our own self-experience testifies to the existence of all three levels in our own consciousness: our experience of our bodies and physiological processes, of our desires and emotions, and of our thoughts. The effort to study the physiological correlates of this self-experience is a useful field of scientific study that can provide insight into the interaction between these three modes of our being. But the insistence on reducing all self-experience to its physiological correlates is not justified by reason, evidence, or self-experience. "If pushed to its extreme, it would give to a stone or a plum pudding, a greater reality and to thought, love, courage, genius, greatness, the human soul and mind facing an obscure and dangerous world and getting mastery over it an inferior dependent reality."†

---

*Sri Aurobindo, *The Life Divine* (Pondicherry, India: Sri Aurobindo Ashram, 1970), 3.
†Sri Aurobindo, *The Life Divine*, 647.

Merely stating an alternative hypothesis does not validate it. Indeed, it raises further questions that require serious consideration. Foremost are the inherent reason and the fundamental process by which an infinite universal consciousness could render itself involved or self-absorbed in apparently inanimate material forms and then progressively evolve out of that self-concealed status to manifest the rising forms of life and consciousness we find in the universe. The purpose, rationale, mechanism, and process of this evolutionary process is examined in depth by Sri Aurobindo.*

Individuality

A glaring inadequacy of modern science has been its dealing with the phenomenon of conscious individuality. For science is the study of types, common characteristics, and unvarying universal laws. It is well suited for the classification of the elements in the periodic table, the classification of biological forms by phyla, genus, and species, and countless other typal phenomena. But when it comes to conscious human beings, the limitations of classification, universal law, mathematical precision, and statistical probability become evident. Nature is regarded by science as an inconscient energy that is entirely impersonal. The emergence of consciousness and the creation of personality appear to it as an inexplicable mystery.

The conscious individual is the pinnacle of the evolution of life and consciousness. The individual is the most complex phenomenon in the known universe. True individuality as described by eminent psychologists such as Jung, Maslow, and Rogers represents the highest stage in the organization of human personality. The individual is also the catalyst for all the advances of the social collective. Creative, pioneering individuals are the source of the new ideas, discoveries, inventions, and organizations that enable society to enhance its power for defense, pro-

---

*Sri Aurobindo, *The Life Divine*, 647.

duction, transport, communication, science, technological innovation, trade, education, governance, social welfare, artistic excellence, and entertainment.

The development of Individuality is the key to human progress in the past, present, and future. All the great discoveries and developments of the past have resulted from the creative, divergent, original inspirations and actions of individual members of the group who dared to think and act differently than others, to discover, invent, and innovate new ideas, tools, technologies, organizations, and activities that have been subsequently adopted by the collective and incorporated into the mainstream of social existence. *To reduce the most remarkable phenomena in the universe to a series of mechanical processes and chance events requires an act of blind faith contrary to our highest reason and self-experience.*

Individuality is the seed from which human aspiration arises initially as a tiny spark, then evolves progressively in strength into a firm will for personal accomplishment and at its highest into a flame for the collective advancement of humanity. Individuality represents the distilled essence of power in each human being for highest accomplishment both for himself and for humanity. *Society is the infinite; the individual is the infinitesimal.* With the right strategy, the individual can tap at any point and release the infinite power of the society.

All our knowledge is incomplete and inadequate unless it reveals to us the process and method for development of this extraordinary creative capacity of consciousness in ourselves and in others. All the achievements of modern science with its reliance on material processes are inadequate to explain the phenomenon of conscious individuality, which is the most intimate and precious attribute of our humanness. Science will come of age and cross the boundaries into a new age of knowledge when it rejects the limitations of the implicit assumptions on which it is currently founded and seeks to openly explore on its own terms the wonders and mysteries of the evolutionary adventure of consciousness.

Our Knowledge of Life

The greatest discoveries of both science and spirituality have been those that reconcile and integrate phenomena that were previously believed to be independent or even contradictory to one another. Thus, Newton reconciled the principles of inertia and movement, Maxwell unified electricity and magnetism, Einstein discovered the formulas relating matter and energy, space and time.[*]

Scientific knowledge of the physical plane, abstract philosophical mental knowledge, and direct inner spiritual experience all have played valuable roles in the evolution of human consciousness. Science confirms the existence of immutable, universal physical laws of Nature. Spirituality affirms the wisdom of universal and eternal values and subjective self-experience. But neither by itself is adequate to provide the mastery and self-mastery we seek for fullness of being in life. Laszlo argues that the universal laws are not merely accidental, but rather expressions of a concealed evolutionary intention. He also argues that these spiritual values are not merely useful ethical principles, but rather expressions of a higher, yet to be realized spiritual consciousness of reality. Humanity needs a knowledge that combines, reconciles, integrates, and transcends them both.

Neither science nor spirituality has given us the complete knowledge needed to fully understand and live life in its vibrant, organic fullness. Spirituality has been so self-absorbed in the experience of fundamental oneness that it often loses sight of the universal manifestations of that oneness. "We perceive that in the Indian ascetic ideal the great Vedantic formula, 'One without a second,' has not often been read sufficiently in the light of that other formula equally imperative, 'All this is the Brahman.'"[†] Similarly, science has been so self-absorbed in the study of material phenomena that it has lost sight of other planes of experience that are equally real and valid and in one sense even more fundamental.

---

[*]Lee Smolin, *The Trouble with Physics,* chapter 4.
[†]Sri Aurobindo, *The Life Divine,* 24.

The attempt of philosophy in its turn to reconcile and synthesize these two poles through ingenious mental constructions of reality has failed to attain either the vivid heights of spiritual experience or the solid concreteness of matter, leaving us instead in a flatland of abstractions.

Perhaps both extremes, materialism and asceticism, have been necessary to explore the opposite polar ends of the full spectrum of reality. The schism between science and spirituality is symptomatic of the division between our inner subjective world of self-experience and the outer physical world in which we live and act. Life is the testing ground on which science and spirituality meet. By life, I refer to the field of conscious experience by which human beings strive to survive, grow, develop, and evolve. We need a knowledge that will enable us to make the right decisions and achieve the right results in all our actions, great and small. We need a knowledge that gives us the right sense of timing, measure, and proportion. We need a knowledge that leads to fullness of inner being and effective power of outer action. The efficacy of our knowledge is not ultimately to be demonstrated in a laboratory or a factory or in the meditations of spiritual contemplation, but in our capacity for mastery in the field of life in which the inner and outer, objective and subjective, material and spiritual meet in our awareness, experience, and actions as conscious beings.

The solution cannot be arrived at by a mental resolution or reconciliation, or at least not by that alone. It requires us to discover the knowledge of the correspondence between our inner consciousness and outer circumstances, which is the hallmark of wisdom. That knowledge reveals to us the great discovery of the direct power of consciousness over life and the means by which changes in our consciousness can result in change in our lives and in the life of the world around us. The conscious individual is the point of reconciliation between matter and spirit and the pioneer of evolving consciousness in the universe. The possession of that knowledge and power will be the climax of humanity's ascent from the animal and the fulfillment of the human aspiration for inner spiritual perfection and perfection in outer life. The destiny of science and spirituality is to achieve a reunification of inner and outer knowledge in a living synthesis. That knowledge will reveal the true character of life depicted by great writers such as Shakespeare and the power to make life respond.

Laszlo provides an answer to two of the most fundamental questions that have preoccupied and perplexed humanity through history. Yet if the prevailing hypotheses of modern science are correct, then any answer to these questions is inconsequential and irrelevant. For if matter and material energy are the fundamental reality and if that reality is solely governed by the twin gods of chance and necessity—as assumed by current evolutionary theory—these questions have no meaning or answer whatsoever. In that case, human beings are merely an agglomeration of material elements formed by accident according to some arbitrary laws and constants of physical nature. Then life is merely a manifestation of electrical and chemical processes and the will to live merely an appearance, for then there is no will in material nature other than chance and necessity. Then consciousness is merely an artifact of material processes, the sense of self and self-awareness a mere illusion, and neither scientist nor scientific knowledge has any real significance or validity. To inquire regarding purpose in this case is meaningless because nothing has any purpose in the universe.

This conclusion is diametrically opposite to our experience as human beings. For will, consciousness, and purposefulness are the very essence of human experience. The will to live, grow, enjoy, discover, know, love, and create has been the highest aspiration of humanity throughout history. Reason, insight, intuition, and self-experience reject this conclusion as fallacious and born of false premises, a too narrow and limited conception of knowledge born of an excessive preoccupation with only the physical dimension of reality and confined to the most physical appearances of nature.

Laszlo presents an alternative hypothesis and an alternative answer based on a broader conception of reality and a more comprehensive and integral conception of knowledge. His hypothesis affirms not only the purposefulness and validity of human experience, but the purposefulness of the universe as the venue for the evolution of individual consciousness by cosmic consciousness. His view affirms our highest aspirations and deepest intuitions regarding the phenomena of life, consciousness, will, intention, and purposefulness—both at the individual and at the universal level.

Laszlo's facts and arguments do not and cannot disprove the material hypothesis any more than physical science can disprove or invalidate spiritual experience. But what they can do is to make us pause and seriously reconsider the fundamental premises on which present knowledge is based, and point the way to the boundaries beyond which science is destined to evolve in its continued quest for answers to the ultimate questions of human existence. Laszlo's essay provides the theoretical foundation for a science of life and human fulfillment, with matter as its building blocks, consciousness as its creative motive power, the individual as the key to human and cosmic evolution, and our aspiration as the ultimate determinant of human destiny.

Afterword

James O'Dea

I once visited a site in Turkey commemorating the wily, populist thirteenth-century philosopher Nasreddin Hodja (also known as Mulla Nasrudin). It was a place with great iron gates secured with an imposing set of locks and chains. Not only was it not inviting, it was apparently impenetrable. Confronted with this impassable obstacle to ingress one might turn around mystified and leave none the wiser . . . were it not for the fact that as one steps back something becomes dramatically obvious: there are no walls, fences, or obstructions on either side of the gates, one simply has to walk around them. Would that some of the great mysteries concerning the nature of the universe, and our place in it, were so easily and playfully resolved. One thing is clear: we need a wideview perspective to answer the most fundamental questions about our true identity and the reason we exist.

While *Intelligence of the Cosmos* does not answer every query you might have on these seminal questions, it does allow you to step back and gain the vital perspective offered by new science and new cosmology. This book offers you templates of meaning that help you circumvent entanglement in the locks and chains of questions that can consume your attention and still remain irresolvable without a wider lens to view their relation to each other.

But classical scientific inquiry has a tendency to zero in on the locks. We try to find the key to the mind-matter lock and end up with a cosmic split between the two or, even worse, with matter so dominant it leads us to propose materialistic solutions to almost everything. Consciousness, from this perspective, gets chained to matter and can only serve it like a horse tethered to a wagon.

But once we understand how consciousness actually designs the wagon, tethers the horse, and serves as the driver, we are able to step back and see its pivotal role in the great chain of cause and effect. This is a crude metaphor but having this understanding encourages us to explore the nature and powers of consciousness and experience greater freedom to make wiser choices. Rest assured, through his brilliant scholarship and vast erudition Ervin Laszlo offers real science not just metaphors or tantalizing conjectures to reveal the centrality of consciousness in the causal weave.

Many of the concepts of conventional science and cosmology remain entangled in dualistic frameworks. For example, spiritual reality may be acknowledged, but since there is considered to be no physical evidence for it, it is placed in a separate reality. Spiritual reality is confined to the sacred mystery—a location that exists beyond the verifiable and observable mechanisms of the universe and the laws that govern matter and its substrates. From this perspective God or Source sets the universe and its laws in motion through its all-powerful creative agency, but then sits outside of its own Creation.

In this scenario, the conscious human must transcend all the limitations of matter, biological impulse, social behavior, cultural constructs, pain, and suffering in a quest to arrive at the truth of the Absolute that is outside existence. But this duality cannot exist in Laszlo's breathtaking exposition of the nature of consciousness and its pervasive interfusion in energy-matter. We see that consciousness is never a separate phenomenon, but everything and every being is an aspect of consciousness unfolding.

As our wide-lens view of consciousness increases, we begin to recognize not only its primary position in the chain of causality but also its interactive aspect in evolutionary selection and adaptation processes. Even though evolution is not linear and sequential, it is extraordinarily creative-adaptive, opportunistic, and purposeful. If evolution were a video game with levels that begin with the convening of the elements and calibrating the conditions for life to begin, proficiency would be about mastering complexity as the game moves from the emergence of single-celled organisms to full-scale, techno-driven planetary civilization where billions of people and diverse cultures could live peacefully and sustainably.

One thing playing such a game would immediately clarify would be the way adaptation strategies for surviving and thriving in variable challenging conditions require both the aspect of continuity as well as the key element of innovation. As it happens, Nature is very good at recording and storing information without which life's progress would be severely restricted and assuredly descend into chaos. But as Laszlo has shown in several books, the Universe itself has an infinite chronicle of its own reality in what he calls the Akashic field. The more we are able to peer into the storehouse of information in the libraries of Nature and the Universe the better we are able to step back and see the grand design of it all. Let's look for a moment at what can happen when we look into the great living library of our DNA. In an effort to contrast the prejudices people have with the facts about who we actually are, travel company Momondo gathered sixty-seven people from various ethnicities and offered them a DNA test. The test subjects had one thing in common: they were proud of their nationality, ethnicity, and heritage. But they held some bias against people from other races, whom they considered different from themselves. When they got the results, they saw to their utter amazement that they carried sets of DNA from some of the very ethnicities they had a problem with. Science can indeed help us step back and get a true perspective on who we are. It can help break down the illusion of separate identities that is the cause of so much intolerance and war resulting from a lack of coherence in our basic understanding of who we are. Not only does DNA science help us see how all human beings are connected, but it provides an astonishingly coherent framework for us to understand how we are connected in the great circle with all life. *The Intelligence of the Cosmos* is an ambitious book. Based on Laszlo's previous books and studies, it presents a worldview that integrates rigorous science with authentic spirituality. The vision it expresses is compelling, but it is also deeply coherent. For the author coherence is a central theme. However, there is an acknowledgment by Laszlo that humanity is now reaching a period of critical incoherence where conflict becomes viral. Reality has become a smorgasbord of options that people pick and choose from without deep scrutiny as to whether their choices pass the test of coherence. The result of this incoherence is that our planet's climate balance and ecosphere are dangerously imperiled, and the social order is gravely polarized.

When it becomes clear that even as we have distinct personal psyches and individual ideas, we see that the truth is actually that we all share a common field of consciousness. This knowledge allows us to transcend incoherent dualities that create splits between mind and matter, as well as between science and spirituality. Equally, when we perceive and understand how consciousness is not made up of fragments but is derived from a common source, we reach a breakthrough understanding that who we are is, in fact, an integral part of Source.

Now as we edge toward a deeper understanding of the continuity of consciousness after physical death, as this book suggests, we can also appreciate the coherent way in which the universe—the Akashic field—conserves its learning through us. To the extent that we transmit our learning, we evolve more coherently: it is not only our ideas that become more coherent, but our systems of governance, and the lived quality of our relationships also reflects greater coherence. We learn through embodiment and action what would otherwise remain abstract. Our responsibility is to share with future generations spiritual insight, intellectual discovery, and the knowledge gained through direct experience.

The universe seems to be designed to help us with this task. A mystic might say the universe is designed for this task.

For the great Sufi master Ibn Al Arabi, the Reality, whom some call God, created the entire cosmos as a mirror so as to know its own essential qualities as reflected in the testing conditions of Creation. Thus the God-Reality wanted to know something that could only be known through an evolutionary learning process. Ibn Al Arabi sees the qualities of the divine Reality as revealing its nature—its beauty and truth—under the testing provided by existence itself.

What will it take for us to collectively reveal and truly manifest the qualities and capacities of the divine Reality here on Earth? One thing is certain, it begins with having a reliable and accurate map of the world's reality. *Intelligence of the Cosmos* is a compelling and deeply coherent map for humanity for the testing road that lies ahead.

Biographical Notes on the Author and Contributors

The Author

Ervin Laszlo spent his childhood in Budapest. He was a celebrated child prodigy, with public appearances from the age of nine. Receiving a Grand Prize at the Geneva International Music Competition, he was allowed to cross the Iron Curtain and begin an international concert career, first in Europe and then in America. At the request of Senator Claude Pepper of Florida, he was awarded U.S. citizenship prior to his twenty-first birthday by an act of Congress.

Laszlo received the Sorbonne's highest degree, the Doctorat ès Lettres et Sciences Humaines, in 1970. Shifting to the life of a scientist and humanist, he lectured and taught at various U.S. universities, including Yale, Princeton, Northwestern, the University of Houston, and the State University of New York. Following his work on modeling the future evolution of world order at Princeton, he was asked to produce a report for the Club of Rome, of which he was a member. In the late 1970s and early 1980s, Laszlo ran global projects at the United Nations Institute for Training and Research at the request of the secretary-general. In the 1990s his research led him to the discovery of the Akashic field, which he has continued to study and expound ever since.

The author, coauthor, or editor of ninety-three books that have appeared in twenty-four languages, Ervin Laszlo also published several hundred papers and articles in scientific journals and popular magazines. His autobiography was published in June 2011 under the title *Simply Genius! And Other Tales from My Life* and a one-hour television special is being produced on his life with the title *Life of a Modern-Day Genius*. He is a member of numerous scientific bodies, including the International Academy of Science, the World Academy of Art and Science, the International Academy of Philosophy of Science, and the International Medici Academy. He was elected a member of the Hungarian Academy of Science in 2010.

Laszlo is the recipient of various honors and awards, including honorary Ph.D.s from the United States, Canada, Finland, and Hungary, an honorary professorship at the Institute of Technology of Buenos Aires and honorary citizenship of the city of Buenos Aires. He was awarded the Goi Peace Prize of Japan in 2001, the Assisi Mandir of Peace Prize in 2006, the Polyhistor Prize of Hungary in 2015, and the Luxembourg Peace Prize of the Shenghen Association in 2017. He was nominated for the Nobel Peace Prize in 2004 and 2005.

Ervin Laszlo is founder and president of the Club of Budapest, chairman of Eternea, Inc., honorary member of the Board of Greenheart International, and founder and director of the Laszlo Institute of New Paradigm Research.

The Contributors

John R. Audette earned a master of science degree from Virginia Tech. A native of South Florida, his professional career spans over three decades in hospital, hospice, and medical administration. He also served in senior management positions in public broadcasting and the performing arts.

He has published articles about spiritually transformative experiences and nonlocal consciousness in several scholarly books and has worked closely with many leading figures in this field of scientific inquiry for over four decades. He is the primary founder of the International Association for Near-Death Studies, Inc. (IANDS.org) with cofounders Raymond Moody, Bruce Greyson, Kenneth Ring, and Michael Sabom. He currently serves as CEO of Eternea, Inc. (eternea.org), which he cofounded with prominent neurosurgeon and bestselling author Eben Alexander and with Edgar Mitchell, the late Apollo 14 astronaut and sixth man to walk on the moon. He is an honorably discharged veteran with over six years of voluntary service in the U.S. Army during the Vietnam era.

Kingsley L. Dennis, Ph.D., is a sociologist, researcher, and writer. He previously worked in the Sociology Department at Lancaster University, UK. He is the author of numerous articles on social futures, technology and new media communications, global affairs, and conscious evolution. Kingsley is the author of several critically acclaimed books, including *The Sacred Revival* (2017), *The Phoenix Generation* (2014), *Breaking the Spell* (2013), *New Revolutions for a Small Planet* (2012), *Struggle for Your Mind* (2012), *New Consciousness for a New World* (2011), *After the Car* (2009), and the celebrated *Dawn of the Akashic Age* (2013 with Ervin Laszlo). He coauthored the study "New Media for a New Future: The Emerging Digital Landscape for a Planetary Society," produced as part of the Fuji Declaration for the Goi Peace Foundation, in collaboration with the Club of Budapest. He currently serves as director of publications for the Laszlo Institute of New Paradigm Research.

Shamik Desai grew up in the San Francisco Bay area and studied economics at Stanford University and Oxford. He worked as a banker at various institutions (Morgan Stanley in New York, Cisco Systems in the Silicon Valley, and the World Bank in Washington, D.C.), which gave him a window into some of the societal challenges pressing on the world today. Following this, he pursued a master's degree in international public policy (with a focus on political philosophy) at Johns Hopkins University and wrote a satirical-philosophical novel. He currently leads the ConsciousWorld project, which seeks to map and encourage conscious evolution in all branches of civilization. He is a contributing writer to various philosophical publications and serves as director of special projects at the Laszlo Institute of New Paradigm Research in Italy.

Jane Goodall, Ph.D., DBE, is a British primatologist, ethologist, anthropologist, and UN Messenger of Peace. Considered to be the world's foremost expert on chimpanzees, she is best known for her fifty-five-year study of social and family interactions of wild chimpanzees in Gombe Stream National Park, Tanzania. She has published fifteen books, ranging from her first, *My Friends the Wild Chimpanzees* (1969), to *Seeds of Hope, Wisdom, and Wonder from the World of Plants* (2013) as well as eleven children's books. Dame Goodall is the founder of the Jane Goodall Institute and the Roots & Shoots program, currently with over 150 chapters in different parts of the world. She has worked extensively on conservation and animal welfare issues and has served on the board of the Nonhuman Rights Project since its founding in 1996.

Garry Jacobs is an American social scientist and business consultant. He is currently vice president of the Mother's Service Society (MSS), a social science research institute in Pondicherry, India. He is chief executive officer of the World Academy of Art & Science (U.S.A.), chairman of the Board of Directors and chief executive officer of the World University Consortium (U.S.A.), distinguished professor of interdisciplinary studies at the Institute for Person-Centered Approach (Italy), member of the Club of Rome International (Switzerland), chair of the Advisory Board of Global Institute for Integral Management Studies (Kerala), and managing editor of *Cadmus Journal.* Since 1972 he has been engaged in research on the application of Sri Aurobindo's thought in the fields of business, economics, education, employment, history, governance, international security, literary criticism, organizational theory, philosophy of knowledge, psychology, social development, and spirituality in life. From 1989 to 1994, he was member-secretary of the International Commission on Peace and Food and editor of the commission's report to the United Nations. He is coauthor of two business books on corporate growth; a book on Indian development; a novel on spirituality, science, and business; and more than one hundred articles on various aspects of development, economics, education, global governance, international security, law, and management.

Dawna Jones speaks, writes, conducts workshops, and works with clients, colleagues, coaches, and emerging leaders. She is the author of *Decision Making for Dummies,* placed on Steve Denning's (*Forbes*) list of eight noteworthy books. Jones is host of the "Insight to Action" podcast for business innovators and is a monthly blogger for Huffington Post Great Workplace Cultures. Her interests include exploring approaches to transforming complex organizations to free human potential and match demands to shorten the cycle of application. She is a regular contributor to professional organizations that aim to restore health and adopt a wider view along with design thinking.

Emanuel Kuntzelman is a philosopher, writer, public speaker, and social entrepreneur. He is the president of Greenheart International, a nonprofit organization he founded in 1985, based in Chicago. Greenheart is one of the leading cultural exchange organizations in the United States, which also has branches devoted to fair trade (the Greenheart Shop) and environmentalism and personal development (Greenheart Transforms). He is also the president of the Foundation for the Future in Madrid, Spain, which he established in 1995 along with his wife, Laura Rose. Emanuel is the initiator of the Global Purpose Movement and has been a lifelong spiritual seeker. His book, *Riding the Wave of Global Purpose,* is forthcoming.

James O'Dea is author of several acclaimed works, including *The Conscious Activist, Cultivating Peace,* and *Soul Awakening Practice* (June 2017). He is a former president of the Institute of Noetic Sciences and served as the Washington office director of Amnesty International and as CEO of the Seva Foundation. O'Dea worked with the Middle East Council of Churches in Beirut during a time of war and massacre and lived in Turkey for five years during civil upheaval and coup d'etat. He has taught peacebuilding to over a thousand students in thirty countries and conducted frontline social healing dialogues around the world. A member of the Evolutionary Leaders group and of the Advisory Board of the Peace Alliance and *Kosmos* journal, O'Dea is an honorary fellow of the Laszlo Institute of New Paradigm Research.

Maria Sagi, Ph.D. began her career as a classical pianist and turned subsequently to the study of medicine and psychology, specializing in personality theory, deep psychology, social psychology, and the psychology of music. She received her doctorate in psychology at the Eötvös Lóránd Science University of Budapest and was subsequently awarded the C.Sc. degree (Candidate of the Hungarian Academy of Science). Maria Sagi has done extensive research on the psychology of art and music, as well as on the Hungarian folk-dance movement, theater, and cultural trends and values. Her books include *Aesthetics and Personality, Culture and Personality,* and *Creativity in Music,* coauthored with Ivan Vitanyi. She is the author and coauthor of eleven books and some 150 articles and research papers on social and personality psychology, the psychology of music and art, and information medicine. A former research associate and scientific secretary of the Institute for Culture in Budapest and research associate of the Sociological Institute of the Academy of Science, she currently serves as science director of the international think tank the Club of Budapest.

Books by Ervin Laszlo
A Selected Bibliography

1963

Essential Society: An Ontological Reconstruction. The Hague: Martinus Nijhoff.

1966

Beyond Scepticism and Realism: An Exploration of Husserlian and Whiteheadian Methods of Inquiry. The Hague: Martinus Nijhoff.

1969

System Structure and Experience: Toward a Scientific Theory of Mind. New York: Gordon & Breach.

1970

La metaphysique de Whitehead: Recherche sur les prolongements anthropologiques. The Hague: Martinus Nijhoff (in French).

1972

Introduction to Systems Philosophy: Toward a New Paradigm of Contemporary Thought. New York: Gordon & Breach; Toronto: Fitzhenry & Whiteside.

The Systems View of the World: The Natural Philosophy of the New Developments in the Sciences. New York: George Braziller; Toronto: Doubleday Canada; Oxford: Basil Blackwell, 1975.

1987

Evolution: The Grand Synthesis. Boston: Shambala New Science Library.

1988

L'ipotesi del campo. Bergamo: Pierluigi Lubrina Editore (in Italian).

1991

Editor. *The New Evolutionary Paradigm.* Keynote volume of General Evolution Studies. New York: Gordon & Breach.

1993

The Creative Cosmos: A Unified Science of Matter, Life, and Mind. Edinburgh: Floris Books.

1995

The Interconnected Universe: Conceptual Foundations of Transdisciplinary Unified Theory. River Edge, N.J.: World Scientific Ltd.

1996

The Whispering Pond: A Personal Guide to the Emerging Vision of Science. Rockport, Mass.: Element Books.

1999

with Stanislav Grof and Peter Russell. *The Consciousness Revolution: A Transatlantic Dialogue.* Boston: Element Books.

2003

The Connectivity Hypothesis: Foundations of an Integral Science of Quantum, Cosmos, Life, and Consciousness. Albany: State University of New York Press.

2004

Science and the Akashic Field: An Integral Theory of Everything. Rochester, Vt.: Inner Traditions.

2006

Science and the Reenchantment of the Cosmos: The Rise of the Integral Vision of Reality. Rochester, Vt.: Inner Traditions.

2008

Quantum Shift in the Global Brain: How the New Scientific Reality Can Change Us and Our World. Rochester, Vt.: Inner Traditions.

with Jude Currivan. *Cosmos: The Co-Creators Guide to the Universe.* New York: Hay House.

2009

The Akashic Experience: Science and the Cosmic Memory Field. Rochester, Vt.: Inner Traditions.

2013

with Kingsley Dennis. *The Dawn of the Akashic Age: New Consciousness, Quantum Resonance, and the Future of the World.* Rochester, Vt.: Inner Traditions.

2014

The Self-Actualizing Cosmos: The Akasha Revolution in Science and Human Consciousness. Rochester, Vt.: Inner Traditions.

with Anthony Peake. *The Immortal Mind: Science and the Continuity of Consciousness beyond the Brain.* Rochester, Vt.: Inner Traditions.

2016

with Larry Dossey and Jean Houston. *What Is Consciousness? Three Sages Lift the Veil.* New York: SelectBooks.

with Alexander Laszlo. *What Is Reality? The New Map of Cosmos and Consciousness.* New York: SelectBooks.

Index